PATIENT
SAFETY

A Human Factors Approach

PATIENT SAFETY

A Human Factors Approach

SIDNEY DEKKER

CRC Press
Taylor & Francis Group
Boca Raton   London   New York

CRC Press is an imprint of the
Taylor & Francis Group, an **informa** business

CRC Press
Taylor & Francis Group
6000 Broken Sound Parkway NW, Suite 300
Boca Raton, FL 33487-2742

© 2011 by Taylor and Francis Group, LLC
CRC Press is an imprint of Taylor & Francis Group, an Informa business

No claim to original U.S. Government works

Printed in the United States of America on acid-free paper
10 9 8 7 6 5 4 3 2 1

International Standard Book Number: 978-1-4398-5225-5 (Paperback)

Visit the Taylor & Francis Web site at
http://www.taylorandfrancis.com

and the CRC Press Web site at
http://www.crcpress.com

Contents

Acknowledgments

I would like to thank all the people in healthcare in places ranging from Australia to Sweden, the Netherlands, Canada, the United Kingdom, and the United States for their willingness to enlighten me about their world and work in healthcare. I am particularly grateful for the insights offered to me by Rob Robson, Tom Hugh, Richard Cook, Isis Amer Wåhlin, Natalie McGregor, and Jo Anne Waterhouse. I also want to thank Darrell Horn for his readings of earlier drafts and numerous improvements.

Preface

Patient safety is seen by many in healthcare as a pressing issue, yet achieving patient safety is difficult. The risks to patients are many and diverse, and the complexity of the healthcare system that delivers them is huge. The human factors approach maintains that the creation of safety cannot be up to a few good doctors or some exceptionally dedicated nurses or technicians. Of course these individuals exist, and they form a fantastic asset to any healthcare system. They may be the champions who push for a preoperative checklist in the hospital, for a time-out, for a new hand-washing basin in a smarter place, or for team training.

But just building an organizational safety strategy on individual heroes is a bad idea. It is brittle. What if the heroes leave or burn out? What if they do not succeed in overcoming the many institutional obstacles that conspire against their ability to create safety? People in healthcare often think that safety lies foremost in the hands through which care ultimately flows to the patient—those who are closest to the patient, whose decisions can mean the difference between life and death, between health and morbidity. That is the point at which we should intervene to make things safer, to tighten practice, to focus attention, to remind people to be careful, and to impose rules and guidelines.

The human factors approach refuses just to lay the responsibility for safety and risk at the feet of people at the sharp end. Instead, the human factors approach looks relentlessly for sources of safety and risk *everywhere* in the system—the designs of devices, the teamwork and coordination between different practitioners, communication across hierarchical and gender boundaries, the cognitive processes of individuals, the organization that surrounds and constrains and empowers them, the economic and human resources offered, the technology available, the political landscape, even the culture of the place.

This book takes a human factors approach to creating patient safety. Chapter 1 inquires after the often strained relationship among competence, standardization, and error in healthcare. Medicine remains strongly preoccupied with the autonomy and discretion of its individual actors. But is competence an individual virtue? Or is it (i.e., its creation, assurance, and maintenance) in large part a responsibility of the system? The second chapter delves into the complexity of the human error problem in healthcare. Error is not a cause of trouble but rather a symptom. And error alone rarely determines success or failure in healthcare. Many organizational, operational, and technological features that make medical work safe or unsafe lie beyond the reach of individual competence.

Chapter 3 covers the cognitive factors of work in healthcare. It discusses attentional dynamics (where to focus attention, when, at the cost of what else), knowledge factors (how to learn, store, activate, and deploy knowledge), and strategic factors (how to deal with larger constraints and goals imposed by the organization in which the work is carried out). These three together can often carry a good deal of the explanatory load for why things go well or wrong. Technology is often seen as one way out of a human error problem. If technology does the work, then humans cannot make errors in doing that work. But as Chapter 4 shows, technological change transforms people's work and their workplace, including many human-to-human relationships. It creates new sources of strength and brittleness.

The human factors approach to patient safety says that the major source of risk lies not with individual caregivers but with the system surrounding and interacting with those caregivers; this is where risk to patients brews and grows and where risk should be most effectively recognized, managed, and contained. Chapter 5 covers the most important schools of thought on safety culture and organizational risk. Chapter 6 offers organizations a concrete set of human factors approaches to managing or containing such risk: safety reporting and organizational learning, adverse event investigations, resource management training, and checklists.

Chapter 7 discusses the difficult balance between learning and accountability, recognizing that attributing adverse events to human error or violations (and sanctioning the human for those) does little to improve the healthcare system. Accountability is often equated with sanctions in healthcare. This is an oversimplification, just as blame-free and accountability-free are not the same. The chapter concludes with some of the causes and consequences of the criminalization of human error in healthcare.

Although complexity is a defining characteristic of healthcare today, many of its managers and practitioners often act as if it were a merely complicated system. Chapter 8 takes on the conflation of complicated and complex systems—a difference that matters enormously. What might work in the former (standardization, best practices, studying component behavior) is all but useless in the latter. Complexity theory offers a new and exciting frontier in our understanding of patient safety, and the book finishes with a discussion of it.

The breadth of the human factors approach to patient safety is itself testimony to the realization that there are no easy answers, no silver bullets. If there were, they would have been found and applied long ago. Of course, locally, in small pockets, some quick progress can probably still be made by applying one or another human factors tool (e.g., checklist, team training). And not considering those opportunities could easily be construed as unethical. But oversimplifying the challenge to address patient safety is perhaps one of the biggest mistakes we can make. In a system that in most countries is increasingly pressed for money and human resources, there are no simple solutions. Often, there are only hard choices. What the human factors approach does do is offer the substance and the guidance to consider them in all their nuance and complexity.

1 Medical Competence and Patient Safety

"If only I could get rid of the nurses who make mistakes, things would be a lot safer around here," the nurse manager sighed.

We were in a large trauma hospital, and like many hospitals, it was becoming aware of the problem of patient safety. How was it going to protect patients from preventable adverse events? The nurse manager had a clear idea about the source of risks for patients: It lay with her unreliable, unsafe nurses who made mistakes. They made mistakes with patient identification and mistakes with dosage and drug administration. They violated rules and routines related to infusion devices. They made mistakes in drug selection and labeling. They made mistakes when informing doctors or family or mistakes in doing whatever else they were asked.

The hospital had not had an easy time recruiting nurses, and the nurse manager might have felt that this led to lower standards. Reliance on agency nurses and nurses from other countries had increased. For the manager, the patient safety problem came down to the inadequacy or the incompetence of some of those nurses. If only she could get rid of the less competent practitioners, things would be safer around her ward.

When things go well, healthcare tends to celebrate "good doctoring"—acts by competent people who succeeded despite the organization and its complexity (Gawande, 2002). Failure, on the other hand, says Gawande, is a result of human ineptitude. When things do not go well (when adverse events occur), healthcare tends to zero in on the people at the sharp end, in direct contact with the safety-critical process, who, for once, failed to hold that complex, pressurized patchwork together—rather than inquire about the systemic sources behind the production of all that complexity.

Nowhere is this simultaneous belief in individual strength and brittleness as persistent as in healthcare. People think that safety lies foremost in the hands through which care ultimately flows to the patient. That is the point at which we should intervene to make things safer, to tighten practice, to focus attention. Thus we can ask caregivers to try harder, to read labels more aggressively, and to double-check more often, with more technology. Or we can get rid of the least safe, most clumsy practitioners at that sharp end and replace them with better ones.

One standard model seems to be that patients can be safer if only we have better doctors, nurses, or lab techs. Of course, competence matters. The way healthcare is organized, there is hardly a good substitute for the medical experience, expertise, and competence of the individual caregiver or for the deference and ethical responsibilities that come with it. There are undoubtedly ways in which we can all do better, in which we can apply more focus and be more conscientious or diligent. Paying attention to what you are doing does matter; it can make a difference, says Gawande (2008). In fact, it is what the ethical, or deontological, commitment of being a caregiver means. *Deontology* refers to the duty, the obligation that is linked to one's profession and relationship to other human beings. Nothing should be more important than the patient who is in your care right then and there. Pellegrino (2004) contended that rather than "systems," healthcare needs individuals with "strength of character to be virtuous." Promoting "systems" thinking undermines the unique fiduciary relationship between caregiver and patient and shortcuts personal control over, and accountability for, clinical outcomes.

But are individual virtue, competence, and strength of character—once attained or demonstrated—the only things, the *main* things, we want to rely on for ensuring that sacred duty? Even healthcare itself wavers on the question. Twenty percent of staff surveyed by Gawande (2010) about a surgical checklist (which, in another study, nearly halved surgical deaths) said the list was not easy to use, and that it did not improve safety. Yet 93% wanted to have the checklist used when they were undergoing surgery. The deontological principle of medicine means that nothing is more sacred than your obligation to the patient in your care—except when you are the patient yourself. Then suddenly that surgical checklist sounds like a good idea. The ethical obligation and fiduciary relationship that form the bedrock of the unique subculture of medicine can apparently no longer be trusted to provide safe care. Individual competence and commitment are not enough: A standardized tool should be used.

COMPETENCE AS INDIVIDUAL VIRTUE OR SYSTEMS ISSUE?

Competence problems come from somewhere. People lack competence to accomplish certain tasks not because they are somehow deficient or defective. Of course, there are individual differences between people. Some people are more suited for a particular profession than others are. The various stages of medical training are themselves a selection device to sort some of this out, weeding people out and distributing them across specializations that might suit them best (Bosk, 2003).

However, there is consensus that medical error problems persist independent of seniority or specialization (Kohn, Corrigan, & Donaldson, 2000). This should get us thinking about competence problems stemming from the features of a much larger system of professional selection, medical education, skills training and maintenance, and proficiency checking. What are the practical possibilities and available budgets for recurrent training, and what amount of time are clinicians given to attend it? Do physicians get necessary credit for updating their clinical skills simply from going to a biennial conference in Barbados? Does new technology just show up one day in the operating room or the ward, accompanied by a manual that nobody will have time to read or the training to really understand?

Among safety-critical fields, healthcare is special in its institutional and cultural assumptions about competence. For example, once a clinician has learned something, there is little agreement on a need to periodically rehearse or check that skill extensively. Also, not many questions might be asked regarding whether the clinician is competent to tackle a task or technology never done or used before (for example, the use of a morcellator in surgery, as discussed in the following example). After all, the clinician already has a once-attained basic qualification and experience.

The Morcellator

A 3-year-old girl died during laparoscopic splenectomy when the operating surgeon applied a new piece of technology called a morcellator, and cut through an artery by mistake. The idea for using the morcellator came up when the team had entered the operating theater and discovered that the morcellator was available in the gynecology theater. "How about using the morcellator?" was the chance remark that set the events in motion. The surgical team had not obtained specific informed consent to use the morcellator, and the operating surgeon received 5 minutes of spontaneous training from the other surgeon, who had seen a morcellator used in action 5 years before. The only surgeon with recent hands-on experience with the morcellator was outside the operating theater, watching this surgery via a remote display of the laparoscopic camera.

What appears on the surface as a simple competence problem often hides a much deeper world of organizational constraints, historical and cultural trajectories, peer and patient expectations and assumptions, and professional dispositions. A simple comparison (see Table 1.1) between two different safety-critical fields, aviation and medicine, shows the most outward signs of vastly different assumptions about competence, its continuity, assurance, and maintenance.

For example, training on a new piece of technology and checking the operator's competency before he or she is allowed to use it in practice is required in aviation. No assumptions are made about the automatic transference of basic flying or operating skills. Similarly, skill maintenance is assured and checked in twice-yearly simulator sessions required to retain the license to fly the airplane. Failing to perform in these sessions is possible and entails additional training and rechecks.

Other aspects of competence, such as the ability to collaborate on a complex task in a team, are tackled using standard communication and phraseology, standard-format briefings before each new operational phase, standard procedures and divisions of labor for accomplishing operational tasks, and the extensive use of checklists to ensure that work has been done. Finally, there are limits on duty time that take into account the inevitable erosion of skills and competence under conditions of fatigue. Novel operations (e.g., a new aircraft fleet or a new destination) are almost always preceded by a risk assessment, by airline and regulator alike, before their approval and launch. Skills, competence, or safety levels demonstrated in one situation are not considered to be automatically transferable to another.

In addition, aviation is capable of generating immediate and almost exhaustive evidence about the effects of any procedural or technological intervention. Modern airliners are equipped with flight data monitoring systems capable of recording hundreds of parameters related to the aircraft and the flight every second. Say, for instance, that the criteria for a stabilized approach are changed. A stabilized approach basically means the airliner has to be at the right speed and configuration

TABLE 1.1

Vastly Different Assumptions about Competency and How to Ensure and Maintain It over Time in Aviation and Medicine

Ideas about Competence	Aviation	Medicine
Type training and check before use of new technology	Always	No
Recurrent training in simulator for skill maintenance	Twice a year	No
Competency checks in simulator	Twice a year	No
Emergency training (both equipment and procedures)	Every year	No
Direct observation and checking of practice	Every year	No
Crew resource management training	Every year	No
Standard-format briefing before any operational phase	Always	No
Standardized communication/phraseology	Yes	No
Standardized procedures for accomplishing tasks	Yes	No
Standardized divisions of labor across team members	Yes	No
Extensive use of checklists	Yes	No
Duty time limitations and fatigue management	Yes	No
Risk assessment before novel operation	Yes	No

Note: The table is a generalization. Different specialties in medicine make different assumptions and investments in competence, and there also is a slow but gradual move toward more proficiency checking, checklist use, and teamwork and communication training in medicine.

and in the right place for making a landing on a runway. The next day, data from thousands of flights will pour into the airline's computers and will be automatically analyzed for the effects of the new procedure compared to immediate history (like yesterday or the same day last week), or perhaps even other airlines. In modern airlines, the saturation rate of flight data monitoring is often close to 99.9%. Basically, all passenger jets are recording these data, all the time.

This gives "evidence-based intervention" a whole new meaning. Contrast this with evidence-based medicine. While the number of randomized clinical trials grew from fewer than 500 per year in 1970 to 15,000 in 2006 (Wachter, 2008), still less than 0.1% of patients are enrolled in these formal clinical studies. Those enrolled are not likely to be representative of the larger patient population, so those in medicine cannot honestly say that they learn much about the effects of a procedural, clinical, or pharmaceutical intervention for the remaining 99.9% (Pronovost & Vohr, 2010). Adherence to so-called evidence-based interventions, then, can generate much weaker clinical outcome results than some would hope (Fonarow et al., 2007).

In aviation, none of these programs, limitations, or checks is perfect either, of course. Gathering recorded flight data is one thing. Weeding through the huge electronic footprints left by each flight in the hunt for patterns or trends requires a completely different level of intelligence, analysis, and synthesis than needed for just harvesting the data. Things are missed, particularly if you do not know what to look for while harvesting, which is where the real dangers can lurk. Also, initial training on a new piece of technology or aircraft has been reduced under economic pressure, sometimes leaving people with only a superficial understanding of the various modes and interrelationships (Federal Aviation Administration, 1996). This is mostly okay, unless things start going wrong, and people are thrown off the beaten path and their routine responses and overlearned heuristics no longer apply. Recurrent training in an airliner simulator does not always help. Pilots are rushed through a large number of tasks and systems to meet regulatory requirements, and there may not be enough time for critical reflection on the work done or the skills displayed. Manual skill erosion that occurs as a result of months or years of work with highly automated technology cannot be fended off by twice-yearly recurrent sessions, and duty time limitations are based more on industrial comprise between unions and management than solid fatigue research.

But the basic assumption remains. In aviation, there is no explicit reliance on competence alone. Competence alone is not trusted to sustain itself or to be sufficient for satisfactory execution of safety-critical tasks. Competence needs help. Competence is not seen as an individual virtue that people either possess or do not possess on release from training and entry into a profession. It is seen as a systems issue, something for which the organization, the regulator, and the individual all bear responsibility for the entire lifetime of the operator.

WHY THE DIFFERENCE IN COMPETENCE ASSUMPTIONS?

Empirical research attested to the large difference in mortality risks to patients versus airline or railway passengers—a difference of orders of magnitude (Amalberti, Auroy, Berwick, & Barach, 2005). In civil aviation, passenger railways and nuclear power plants in Europe, for example, the rate of catastrophic accidents per exposure (e.g., a flight) is better than 10^{-6}. In other words, chances are 1 in 1 million that this flight might end up catastrophically. Healthcare is generally agreed to be no better than 10^{-4}, a figure that varies widely among specialties (anesthesia and blood transfusion have better rates; many forms of surgery have worse). Of course, agreement on such numbers is difficult because of the negotiability of what constitutes an unnecessary death in healthcare (or a death due to an adverse event). But that does not mean that such differences between specialties and even safety-critical fields are not at least instructive.

Patients are basically exposed to three kinds of risk: their disease, the diagnosis and treatment plan, and implementation of that plan. Various specialties also have different encounters of various depths with these various stages. This is different from airline or railway passengers, who (it is hoped) do not have to board a diseased aircraft or train and for whom plans and implementation can be more standardized. In healthcare, this diversity has sustained what Amalberti called an "excessive autonomy" of actors (Amalberti et al., 2005). Dramatic improvements in safety, in fields that do not have as much diversity, have been correlated with a gradual reduction in actor autonomy—particularly through standardization and procedures (Amalberti, 2001). The disinclination for this in healthcare has gone hand in hand with the idea that physicians are unique craftspeople whose exercise of skill is about situational insight, deftness, contextual sensitivity, mastery, and prowess. No doubt, sustaining this position everywhere might have its problems. Actors in medicine are not seen as equivalent—not by peers, not by nurses, not by administrators, not by patients. Some are seen as better, as defter, than others. Amalberti and colleagues (2005) concluded the following:

> Health care professionals must face a very difficult transition: abandoning their status and self-image as craftsmen and instead adopting a position that values equivalence among their ranks. For example, a commercial airline passenger usually neither knows nor cares who the pilot or the copilot flying their plane is; a last-minute change of captain is not a concern to passengers, as people have grown accustomed to the notion that all pilots are, to an excellent approximation, equivalent to one another in their skills. Patients have a similar attitude toward anesthesiologists when they face surgery. In both cases, the practice is highly standardized, and the professionals involved have, in essence, renounced their individuality in the service of a reliable standard of excellent care. They sell a service instead of an individual identity. As a consequence, the risk for catastrophic death in healthy patients (American Society of Anesthesiologists risk category 1 or 2) undergoing anesthesia is very low—close to 10^{-6} per anesthetic episode. (p. 759)

One argument that preserves actor autonomy, and that helps keep a craftsman attitude alive, regards the extraordinary complexity and diversity of clinical problems that can be encountered. This makes formal training on and checks of each new problem (structural-anatomical, physiological, technical) impractical. This argument, however, is not always sustainable. "The sort of surgery I do is incredibly routine," said one surgeon. "Unless there is a problem, I do not change my routine" (R. McDonald, Waring, & Harrison, 2006). In addition, some questions in surgery or medication delivery (such as whether this is the correct patient, correct procedure, correct site, or correct drug) are never going to change, no matter how deviant a patient's anatomy or physiology.

There is another side effect of the craftsman attitude. As compared to other safety-critical fields, healthcare does not extensively regulate its own production demands or set limits on its maximum performance. It seems as if there is always the next patient and more after that. A level of production needs to be obtained, no matter what it takes. Accepting limits on maximum performance is something that is both tightly regulated and culturally institutionalized in many worlds and professions. Imagine an airline crew who have just landed in Hong Kong from London (a 12-hour flight and 7-hour time change) offering to fly back to London right away—just because there is a load of passengers waiting, and the new crew has not shown up yet. The feat could be accomplished within the length of a shift that healthcare practitioners regularly have. Yet it would be illegal, against company rules, and professionally unimaginable—and most passengers would not trust the crew with the job.

Eighteen Bypass Operations per Day

Robert Wachter (2008) reported how Dr. Michael DeBakey, the legendary Texas heart surgeon, once performed 18 cardiac bypass operations in a single day. The outcomes of the patients (particularly patient 18) were not reported, but the subtext of DeBakey's celebrated feat was clear. Real doctors do not complain about their workload. Their endurance can be exploited as a sign of their competence, of their superhuman status and indispensability as skilled practitioners, of their unwavering deontological commitment. They live up to the sacred call to heal and save lives.

There is little research to rebut this kind of virtually limitless production. One problem has been to generate persuasive data on clinical performance impairments. Prospective participants are physicians and nurses who are chronically sleep deprived (which would seriously muddle a study's control and test condition) and have little free time (Gander, Merry, Millar, & Weller, 2000; Sugden, Aggarwal, & Darzi, 2010). In other fields, though, 24 hours of sustained wakefulness has been shown to result in performance decrements equivalent to a blood alcohol level of 0.1% (which is considered legally drunk in the United States and many other countries) (Campbell & Bagshaw, 1991). You cannot safely drive your car home, but you can care for seven patients in the intensive care unit (ICU). Sleeping 5 hours per night for a week produces the same effects (and for many residents, these two disruptions actually combine within a single week) (Wachter, 2008). Yet physicians are ill-calibrated regarding the effects of sleep deprivation on their performance, with the majority saying it has no or little effect on them (Helmreich, 2000).

Another argument against taking a different perspective on competence is the rapid pace of technological development in many medical domains. This regularly offers new treatment opportunities and techniques and creates patient expectations before they can be solidly implemented in medical training. Such evolution and complexity, however, could just as easily be used as strong arguments *for* the development and proficiency-checking of competence. It can also be used as an argument for the introduction of at least some forms of standardization of practice (e.g., cross-team communication) to provide a recognizable basis that is common even in new procedures or techniques.

GOOD DOCTORING AND THE PURSUIT OF PERFECTION

Healthcare has a unique and complicated relationship with issues of standardization and competence. There is a strong preoccupation with the autonomy and discretion of how its individual actors express excellence as well as mediocrity. Individual human virtue is legitimately seen as the basis for safety, and human incompetence as the source of risk (Pellegrino, 2004). Good doctoring creates success; human ineptitude creates failure (Gawande, 2002). Healthcare is quick to see erratic, unreliable people and their incompetence as a meaningful target for intervention, just as it sees heroes and celebrated practitioners in those who overcome clinical and organizational odds to produce successful outcomes *despite* everything and everyone else.

This simultaneous belief in human strength and weakness sets medicine apart from most other safety-critical worlds. Other worlds have seemed more willing to embrace the notion that error and expertise are not only similar but also that they are the joint product of system and individual. There, a belief has developed and become institutionalized that it takes teamwork, or an entire organization, to succeed and that it takes teamwork, or an entire organization, to fail. This has been dubbed *systems thinking* and is fundamental for a human factors approach to safety in any domain. Many in healthcare resist this notion. To invoke "systems problems," it might be thought, is to engage in a "dry language of structures, not people" (Gawande, 2002, p. 73).

There might be something path dependent here, a historical residue of assumptions about medical error, safety, deontology, and the unique status of healers in society. These issues deserve some special attention, particularly against the background of the growing complexity and modernization of care across the Western world.

FROM SACRED STATUS TO CONTRACTUAL DUTY

The modernization of Western societies was accompanied by an evolution from so-called status-determined duties to contractual or commercially determined duties (DeVille, 2004). In a status-determined social relationship, members of society derive their rights, their duties, and their liabilities from their calling, their role, and their status. This would be assumed by a king, for example, a pope, a medicine man or witch doctor (Outram, 2005; Stewart & Strathern, 2004). In many instances, such status and the resulting rights and duties were seen as the result of divine intervention in the lives of exceptional individuals. They were God-given to those who deserved or could live up to them. The entire notion of a "calling" explicitly affirms such external involvement in earthly or personal matters: There has to be some agent or entity who calls. A calling and a status can turn into entitlement. Privileges that were bestowed on a particular person through metaphysical intervention could hardly become the negotiable subject of human squabble.

Enlightenment, modernization, concomitant democratization, and societal maturation started eroding such entitlements. Instead, rights, duties, and liabilities were made to arise increasingly from explicit, conscious agreements of individuals. These were contracts that legally bound the actions and expectations of people. A contract, or other kind of agreement, gives people something to fall back on, to arbitrate outside any calling, status, or rank. A contract in principle rules out privilege or entitlement as it presupposes equality before the law (Cohen-Charash & Spector, 2001; Rawls, 2003).

Not all professions evolved equally quickly, however. The Roman Catholic priesthood, for example, has remained status determined well into the twenty-first century. Only in the wake of its worst spate of pedophile scandals in 2010 did the Vatican direct its bishops to contact the police on reports of alleged child abuse (Bradley Hagerty, 2010). In other words, it called explicitly on its bishops to begin seeing their priests' relationships with parishioners as a contract, an agreement whose breach is supposedly liable, like any other between people in modern society, to judicial scrutiny and legal recourse. Child abuse reports were previously kept largely in-house, stashed away as the secrets of a specially privileged status-determined profession, and critics said that the Vatican statement would do little to change it (Bradley Hagerty, 2010).

Prehistorical and historical accounts of healers in premodern societies, as well as contemporary accounts from native or first-nation communities in a number of countries (Arvigo, Epstein, & Yaquinto, 1994; Conley, 2005; Mulcahy, 2001; Pilz, 1988; Young, Ingram, & Swartz, 1989), stressed the unique position of the healer in their respective societies. Interlocution with the divine or the metaphysical was, and is, a basic precondition for deserving and sustaining the calling and its status. The ability to rule over life and death is one of the rights emanating from it; societies were willing to accept and legitimate that power. Important parts of this image have survived the transformations that perfused Western society since Enlightenment. The practice of medicine is often still seen as a calling to a duty outside normal, secular contractual obligations, a profession of healing and helping (Davis-Floyd & St. John, 1998; Pellegrino, Thomasma, & Kissell, 2000). It is seen by many as a vocation, not work (R. McDonald et al., 2006).

Medicine has retained a unique subculture with its own rules, norms, mythologies, social structures, hierarchies, clothing, tools, and other markers of status, and specialization and identity. Nurses of various types and status, as well as medical apprentices, lab techs, and a range of other professionals, are scattered throughout the middle. The gender division in healthcare is more skewed than in many other professional spheres. There were, for example, essentially no female doctors before the last decade of the nineteenth century. And even today, 92% of neurosurgeons, 90% of thorax surgeons, and 80% of cardiologists are male. The social recruitment of medicine into the various professions and layers of the competence hierarchy directly reflects existing status and power arrangements in society—whether racial, ethnic, gendered, or socioeconomical. Parts of medicine have much in common with the professional and artisan guilds of the seventeenth century, with their tight access controls, rankings into masters and apprentices, and focus on individual artistic expression and discretion. Although capable of production, these are not arrangements built for efficiency and the delivery of reliable quality that started to be demanded of industrial systems from the early nineteenth century onward.

As modernization and industrialization spread across professions in the West, particularly during the nineteenth century, physicians resisted the notion that their relationship with patients was anything like a commercial or contractual one, even if the patients were the ones paying for the service. Worthington Hooker, an influential U.S. doctor in the mid-nineteenth century, commented in 1849 (quoted in DeVille, 2004):

> The relation of a physician to his employers is not shut up within the narrow limits of mere pecuniary considerations. There is a sacredness in it, which should forbid its being subjected to the changes incident to the common relations of trade and commerce among men. (p. 156)

The sacredness in the relationship again implies a calling, the metaphysical intervention in imbuing some members of society with the status that is denied others. The doctor's duties to heal were not based on a contract (or the fact that money exchanged hands) but on the calling and the status. Such duties could never be encapsulated in a contract. John Ordronaux, a contemporary of Hooker (DeVille, 2004), confirmed how "the very nature of the relation between patron and client raised the doctor-patient relationship above all taint of a mercenary character," residing instead in "the character publicly assumed by him who undertakes to render such services" (p. 156).

The judiciary seemed to agree. The unrelenting image of a special doctor-patient relationship that was different from all other commercial, or "mercenary," agreements was one important reason for the development of tort law. It offered a separate arena of legal recourse for medical cases in particular. This was different from how any commercial breach of contract was handled. In fact, medical malpractice was (and still is) not asserted for breach of contract. Doctors do not typically get sued for not doing what they agreed to do but rather for not doing what they were supposed to do. Medical malpractice is asserted when the defendant did things in a manner that was careless, negligent, or not skillful—the sorts of things a layperson should not expect from a status-imbued profession.

The Symbolic Importance of Medical Competence to This Day

The preindustrial, almost magical nature of medical competence, or its "mythic nobility" (Miles, 2004, p. 2), is reproduced structurally today. Take the superhuman working hours that doctors are supposed to be able to meaningfully work, despite studies that show the effects of fatigue. Patients do measurably worse when nurses work shifts longer than 12 hours (Rogers, Hwang, Scott, Aiken, & Dinges, 2004), and ICU residents make fewer errors when they work shifts averaging 16 hours instead of the traditional 36 hours (Landrigan et al., 2004). But attempts to regulate resident working hours run up against all kinds of cultural and institutional walls. Residents are those who do most of the day-to-day doctoring, particularly at night. They stay at the hospital (or stay on call) to care for patients when attending physicians (more senior doctors) go home. The fact that many residents actually used to live at hospitals (and for all intents and purposes, some still do) gave them the labels "house officers" or "interns"—those internal to the house or building, kept inside of its walls.

Even recently, some residents were on call every other night, accumulating 120 hours of work per week (Wachter, 2008). That leaves 48 hours in that week for doing non-work activities (like sleeping). Workweeks of 100 hours were more common, and rules have now been adopted to bring that down to around 80 hours per week. These rules, however, are routinely violated, again making the notion of a "normal" resident workweek negotiable (see Chapter 5). However, invoking a language of healers, ontological commitment, and sacred duty, an editorial in the *New England Journal of Medicine* defended the superhuman working hours. Sure, it said, the long workweeks

> have come with a cost, but they have allowed trainees to learn how the disease process modifies patients' lives and how they cope with illness. Long hours have also taught a central professional lesson about personal responsibility to one's patients, above and beyond work schedules and personal plans. Whether this method arose by design or was the fortuitous byproduct of an arduous training program designed primarily for economic reasons is not the point. Limits on hours on call will disrupt one of the ways

we've taught young physicians these critical values. … We risk exchanging our sleep-deprived healers for a cadre of wide-awake technicians. (Drazen & Epstein, 2002)

The idea that doctors, just like other professionals in modern society, are contract-bound to deliver their service and expertise is explicitly rejected here. The alternative is scoffed. Limits on duty time would turn the status-bound and sacred-call heeding healers into mere "technicians" (even with the putative benefit of them being wide-awake technicians). These healers, after all, have critical values to learn. The ontological commitment, the duty to and responsibility for one's patients, should trump all else.

The question, however, is whether that deontological commitment to the patient is not better served by being rested when clinical decisions have to get made or when a developing emergency calls for split-second interventions. Of course, shorter shift regimes increase the number of handovers and coordination overhead and risk, further depersonalizing the delivery of healthcare. But even healthcare workers themselves may prefer a wide-awake technician at their bedside to a sleep-deprived healer in the 30th hour of his or her shift.

In fact, depersonalization is being offset by other developments in healthcare delivery and medical training. These aim to reimport empathy in new ways and ensure the "healer's" understanding of the continuity and holism of patient and problem. At various medical schools, students are being taught the importance of narrative. Narrative has always held a central role in medicine. Patients typically offer an account of their symptoms in narrative form; physicians tell stories to colleagues during rounds (Berlinger, 2005; Miles, 2004).

At Columbia University, medical students are discouraged from asking diagnostic queries like a computer and to supplant questions such as "tell me where it hurts" with "tell me about your life." It allows them to form an understanding of the context in which disease appeared. Human memory is made for stories, so listening to patient narratives (even during shorter shifts) is an alternative route to the continuity and longitudinal commitment that was once ensured by being a constantly present house officer or resident. Some hospitals encourage physicians to keep two kinds of patient journals. In one, they record clinical observations and values, as they usually would. In the other, they record narratives: stories from and about the patient, and their own thoughts and feelings about these (Pink, 2005). Such narrative competence, or narrative medicine, is not a replacement for technical knowledge and skills. It is a way of delivering such skills much more effectively and humanely.

That said, depersonalized medicine is not just a scientific-bureaucratic imposition from a faceless organization that needs to deliver healthcare efficiently and on budget. It is, of course, also a psychological defense mechanism generated by practitioners themselves. Some would argue that getting too involved in patient narratives can get in the way of getting the job done. It takes large amounts of energy to overcome losses following a closer attachment through narrative exploration. This can add up, particularly in fields in which patient prognoses are generally not good. And hospitals often still value the meeting of production goals over emotional involvement with patients as holistic, story-carrying human beings.

The cottage industry production of some clinical computer technology, and the often-limited usability testing before it gets fielded (Cacciabue & Vella, 2010; Welch, 1998; Yentis, 2010), is another structural marker for the symbolic appeal of medical competence. Human factors and ergonomics principles and techniques ensure the usability of technology through field studies to understand practice, device design, support of procurement decisions, and evaluation of prototypes, as well as making improvements and refinements. The systematic use of such techniques and principles in healthcare is lagging behind other fields. There are few meaningful avenues to report design or usability problems once technology is fielded, and there may be some attraction in the idea that heroic acts and creative workarounds are necessary to make the technology work in practice (Wears & Perry, 2002).

Also, as a matter of routine, the healthcare system accepts deviations from protocol, guidelines, or any other standard—as long as these emanate from its physicians. Handwritten prescriptions survive and potentially create havoc to this day (Dunn & Wolfe, 2002). Handwriting has been shown to lead to far more prescription and medication errors than get reported to regulatory authorities or professional indemnity insurers. Such errors seem to be accepted as a natural part of practice (Peterson, Wu, & Bergin, 1999).

Expensive Illegibility

A U.S. jury found a Texas cardiologist and pharmacist negligent because of the doctor's poor handwriting on a prescription that caused a man's death. The poor handwriting allegedly led to an error by the pharmacist in dispensing the prescription. The cardiologist wrote a prescription that called for the patient (Peterson et al., 1999) to take 20 mg of Isordil (isosorbide) every 6 hours. However, the illegibility of the prescription caused a pharmacist to dispense the same dosage of Plendil (felodipine), although the maximum daily dose was only 10 mg. The doctor and pharmacist were found equally liable and ordered to pay $225,000. (Hirshhorn, 2000)

One study on handwriting and drugs showed that prescription errors account for 70% of adverse medication events (Velo Giampaolo & Minuz, 2009). Inaccuracy in writing, poor legibility of handwriting, the use of abbreviations, or incomplete writing of a prescription (e.g., omitting the total volume of solvent and duration of a drug infusion) can lead to misinterpretation by healthcare personnel. Interestingly, most prescriptions in hospitals are issued by junior doctors (Velo Giampaolo & Minuz, 2009). Although their handwritten directives may have some speed advantages, a better explanation for the survival of handwritten prescriptions is that they perform important social, psychological, and symbolic work for those on the way up in the medical competence hierarchy. Handwritten prescriptions serve as an important confirmation of individual knowledge and status of the doctor. They express the identity of a sage whose hand has issued squiggles to be interpreted by the lesser-status laity down the competence hierarchy.

   The medical competence hierarchy is also reproduced in how curing and caring are divided up along gender and status. Here is a typical, if not stereotypical, way of looking at the organization of healthcare:

- The process of *curing* sits at the top of the competence hierarchy. This is where diagnostics takes place—where symptoms get matched against possible diseases and decisions about treatment plans, interventions, and care are made. Decisions about the continuation of care are made, the ruling over life and death. Medication orders and prescriptions flow mostly from this level, and activities associated with it direct, control, manage, and order. People here are mostly male and work closely with technology. Their patients are typically asleep or anesthetized and can sometimes be regarded as little more than an anonymous carrier of a medical problem that is either interesting or not interesting (i.e., routine). This level sees not only the highest status, rewards, and pay but also the longest hours, work time, and training. There is a sense of inner circle, in which medical expertise gets to rule and decide.
- The process of *nursing* sits in the middle of the medical competence hierarchy. This is where continuity of care across interventions and physician visits is provided, and responsibility for patient well-being is mostly located in this area of the hierarchy. Here, medication orders are received and carried out by pharmacists, blood bank techs, and nurses. Although this is now the subject of crew resource management and other nontechnical training (see Chapter 6), "nurses are taught that if a doctor writes an order, you do it" (Pronovost & Vohr, 2010, p. 109). For their work, there is a limited amount of medical technology (e.g., for medication delivery), and patients are often awake and less anonymous; that is, they are seen as more than the mere carrier of a medical problem.
- The process of *caring* resides at the bottom of the medical competence hierarchy. This is the place where the lowest-status workers of the healthcare system get to deal with the aftermath of curing and nursing. Workers provide bed care and rehabilitation to patients. They may do some limited testing, and they will clean up after the other processes: taking care of wounds and sutures and feeding patients. Medical competence is not assumed here, and the process of caring generally has a low status and low pay. It is populated mostly by women and often racial or ethnic minorities.

Boundary markers between these different levels of the competency hierarchy are typically heavily patrolled. Sweden recently proposed that doctors should no longer wear their white coats because of the amount of microbial flora they carry, thereby increasing the risk of nosocomially infecting patients (Wong, Nye, & Hollis, 1991). An outcry followed in defense of the white coat, often under the pretexts of professionalism, patient expectations, and practicality. Some doctors, for example, argued that they do not have private offices in the hospitals where they work, and that their white coats function as their mobile offices where everything has a place, from pager to pens to phone to stethoscope. But the true color of the debate became more evident when one physician argued, "We don't want to look like nurses" (Tammelin, 2007). Also, structural changes that may imply less rigidity and possible movement through the medical competence hierarchy are often the subject of heated debate. Some see great advantages with the empowerment of nurses, for instance (Pronovost & Vohr, 2010) where others see threats, reduction in quality, and worsening of standards (Vogel, 2010).

Beyond Bags, Beds, and Bedpans

The *Canadian Medical Association Journal* reported that while England ponders the consequences of its recent move to make nursing a university-degree profession by 2013, the impact of Canada's decade-long foray into elevated nursing credentials remains unclear. English unions, healthcare professionals, and patients alike have voiced fears that upgrading entry credentials will lead to less hands-on training, shortages in nursing supply, and a general attitude of being "too posh to wash" among academic nurses. They might be ill-prepared for the menial tasks required in the direct provision of healthcare.

There is also another side: According to the Canadian Nurses Association, "the knowledge, skills and personal attributes that today's health system demands of its registered nurses can only be gained through broad-based baccalaureate nursing programs." The association also cited a study that indicated that there was a 5% decrease in risk of patient death for every 10% increase in the proportion of hospital registered nurses holding degrees.

Also, increased chances for mobility were seen as good. Nurses are still defined by what they do, and differences in how they were educated come out in the wash once they are actually in the workplace. Canadian Institute of Health Information data indicated that 86.5% of Canada's registered nurses with a degree were still working in direct care as of 2006. That suggests that the roles of nurses have not significantly changed, despite the new requirement for credentials.

"In the 1980s, my physician told me I had to get out of nursing, and I remember someone telling me—'You can't do anything else, you're a nurse,'" one nurse commented. "The difference a degree makes is that it opens doors beyond bags, beds and bedpans." However, Lewis worried that as a degree opens doors for some nurses, it may close them for others. "By requiring a baccalaureate, right off the bat you exclude people from lower income brackets, but you also see a double barrel effect because the ongoing cost and stressors of a degree program not only prevent some people from entering but also weed out others along the way."

This indeed could go both ways. The Canadian Nurses Association projected that the shortage of nurses who provide direct clinical care will climb to the equivalent of almost 60,000 fulltime nurses by 2022 (Vogel, 2010).

Two further implications of the symbolic nature of medical competence are worth consideration here. The first is that variation in practice and a dislike of constraints on personal discretion are likely connected more to identity than to cognitive or social limitations of the individual. This has implications for the eventual success of any kind of solution that follows from a human factors approach to patient safety. This does not have equal import throughout the medical competence hierarchy, of course. Unsurprisingly, the aversion against rule-based approaches grows the closer you get to the top of the hierarchy. Doctors are much less likely, for example, to frown on or report violations of clinical protocols by fellow practitioners than nurses or midwifes (Parker & Lawton, 2000). The second is the expectation of perfection in medicine and the legitimacy and celebration of striving for it (Gawande, 2008). If a medical profession is more calling than contract, if it implies a sacred duty, then this in principle guarantees flawlessness and excludes imperfection. This has interesting and paradoxical links to the accelerating evolution of medical technology and success at clinical intervention and thus to what gets categorized as medical "error" today. These two topics are discussed in the remainder of this chapter.

STANDARDIZATION AND THE FEAR OF SCIENTIFIC-BUREAUCRATIC MEDICINE

One of the fears often expressed by clinicians is the imposition of so-called scientific-bureaucratic medicine (R. McDonald et al., 2006). A proliferation of guidelines, protocols, procedures, checklists, and managerial involvement would put shackles on the discretion and competence of individual medical practitioners. It would downgrade the value placed on their knowledge and attempt to force the complexity and diversity of symptoms, diseases, and interventions into context-insensitive tools that can never be a substitute for the sensitivity and experience of real practitioners. Appealing to the special nature of clinical judgment, one physician remarked: "When I make a decision, I have no idea how I do it. More often than not it just comes to me that such and such a thing is the right thing to do" (R. McDonald et al., 2006, p. 188).

This, of course, not only is an issue of identity but also is consistent with research results on the nature of expertise (Farrington-Darby & Wilson, 2006). Much of the knowledge that experts draw on is tacit, submerged in huge mental libraries of scripts and schemata with which practitioners apparently effortlessly match complexes of cues with potential solutions (Klein, 1993). Yet a key component in the reflection by the physician is the ability to practice without guidelines, which places medical work beyond the reach of written rules and externally imposed scripts. One physician commented how "there are guidelines for care, but it's perfectly legitimate to depart from them where there are justifiable reasons" (R. McDonald et al., 2006, p. 188). This holds as long as it is the—or a—physician who gets to decide when and why to depart from them.

The physician might be in good company. Some of the safest complex, dynamic work occurs without procedures. Rochlin and colleagues studied the introduction of ever heavier and more capable aircraft onto naval aircraft carriers (Rochlin, LaPorte, & Roberts, 1987) and noted that

> There were no books on the integration of this new hardware into existing routines and no other place to practice it but at sea. ... Moreover, little of the process was written down, so that the ship in operation is the only reliable manual. [Work is] neither standardized across ships nor, in fact, written down systematically and formally anywhere. (p. 79)

Yet naval aircraft carriers, with inherently high-risk operations, have a remarkable safety record, like other so-called high-reliability organizations (see Chapter 5).

There is growing evidence that clinicians pay lip service to guidelines but resist them in practice (R. McDonald & Harrison, 2004), often to the frustration of managers and administrators. In one study, managers blamed the lack of adherence on surgeons and other doctors in particular, who did not communicate well, who took it upon themselves to make shortcuts, and who were arrogant and uncaring. Physicians were accused of not seeing themselves as part of "the team" (R. McDonald et al., 2006), which putatively explained their limited willingness to engage the guidelines and rules put forward by their hospital.

This limited willingness, however, is not without reason. It has powerful historical precedents. The Apollo program was designed by NASA (the U.S. National Aeronautics and Space Administration) to put a human on the moon and promote manned space flight. In its initial years, the program was characterized by professional accountability. Managers and administrators deferred without much problem to employees and practitioners with technical expertise, to the skills of those at the bottom of the organization. People there had considerable autonomy and power to decide on design directions and necessary testing or prudence. The program was built on trust and respect for the knowledge and abilities of technical experts. Centralization and standardization was used only to help implement new programs. As soon as these were in place, technical expertise again took the upper hand (Vaughan, 1996).

This changed dramatically during the late 1970s and 1980s. Under the pressure of increasingly limited resources, budget pressures, and political accounting, NASA shifted from professional accountability to bureaucratic accountability. Control over design and operational decisions became concentrated at the top rather than being distributed across the bottom of the organization. A stringent system of superior-subordinate relationships was implemented and enforced in organizational configurations and rules. Hierarchical reporting relations began to dominate the technical culture at NASA. Many engineers were shifted to managerial jobs in which they had to supervise or oversee the work done by far-flung contractors. They spent less time in the lab and more in the office or traveling (Vaughan, 1996). Bureaucratic accountability meant that guidelines, regulations, and procedures had become a substitute for professional judgment and technical expertise. Social control became a matter of rank and rule, not peer and profession. Vaughan observed:

> Perhaps the most troubling irony of social control demonstrated by this case is the rules themselves can have unintended effects on system complexity and thus, mistake. The number of guidelines—and conformity to them—may increase risk by giving the false sense of security. But in addition, a proliferation of rules regulating an industry, a task, or information exchange may create confusion, defy mastery, or result in some regulations being selectively ignored. Also, large numbers of rules and procedures create monitoring difficulties for safety regulators, increasing the workload and reducing the possibility of detecting problems. Even the characteristics of the rules themselves can undermine their ability to regulate. (p. 420)

In healthcare today, managers and administrators might see (or at least attempt to sell) standardization, guidance, rules, and regulations as a sign of modernization, as a necessary transformation of their organization, the same way that NASA saw its conversion to bureaucratic accountability as a necessary step toward being a responsible user of tax money. Systems built on bureaucratic accountability will try to tinker from the top down with the medical-labor environment to make it more efficient and increase the quality of its products. The aim is a systematization of processes, all of it putatively "evidence based." But physicians are likely to see something else altogether: an assault on their identities. NASA engineers who were promoted out of their jobs and never again whipped out a slide rule or touched a mechanical system would have experienced the same thing.

The Computerized Operating Room Scheduling System

One study found a group of managers who revered their computerized operating room scheduling system, which had replaced the old paper system (R. McDonald et al., 2006). Managers and administrators were now able to gather the surgical operating lists, including patient and procedure information, from each of the surgical departments. Bureaucratic accountability of the usage of surgical resources was tightened, in other words. A management meeting would happen a week in advance of the planned list to review the planned timetable and make the necessary technical provisions for the operations to be performed. Working arrangements proposed by surgeons and the availability of anesthesia staff would be questioned and clarified. All of this would be updated on the computer system after the meeting and circulated in advance of the list.

Managers were convinced of the crucial role of their system in maintaining order and reducing waste of resources (both human and otherwise). Only a few managers had access to make changes to the system. In fact, if more people would have access to it (particularly doctors), more trouble would be created. The computerized list was a meticulous plan for how to optimally use the operating rooms and their staff.

Despite this, numerous things intervened and interfered to make the lists and computerized predictions quickly outdated and useless. These included delays, staff shortages, or sudden problems in the availability of equipment. Of course, managers duly recorded all these deviations, tried to figure out their causes, and fed all of this back into the computerized system. Printouts of all these deviations from the plan were then hung on walls around the various surgical departments. The deviations and delays were mostly blamed on tardy surgical and anesthetic staff, and the printouts were ignored entirely by clinicians. No meaningful action ever resulted.

Nevertheless, managers were convinced that the system was doing a good job in keeping everybody on track and optimizing resource usage. Even though they could not say how, they maintained that all the documentation created by them somehow went back to the relevant and responsible people, and that it would all help the system get better. The more pressed for evidence, the vaguer and more defensive their narrative became.

The conversion from professional accountability to bureaucratic accountability is typically driven by changes in how the work is funded and how leaders in the organization are in turn held accountable for their decisions. The organizational focus can shift from overcoming technical barriers internal to the profession to overcoming external barriers that may threaten organizational budgets or survival. Depending on the way healthcare is funded, such external barriers are often political. A business ideology gets introduced, and deal making with insurers or politicians can become the most critical administrative activity. Scarcity of resources hollows out both the influence and inherent nature of a technical culture. And outwardly visible signs of supposed waste and inefficiency (long surgical waiting lists, for example, or empty beds) can be taken by managers or administrators as a call to fulfill political promises or expectations.

By projecting and living up to a cultural image and political expectation of modernity, bureaucratic control, and efficiency, administrators attain legitimacy for their decisions even if these lack anchoring in the daily details of technical practice. A gap can thus grow between production goals and real system capabilities (Vaughan, 1996). Vendors of all kinds of computer-based technology can jump in with putative answers to the need for greater efficiency. There is no evidence, however, that such solutions consistently contribute to greater efficiency or patient safety (Nemeth, Nunnally, O'Connor, Klock, & Cook, 2005).

GETTING THE JOB DONE DESPITE THE ORGANIZATION

The gap between the image projected by administrators of a healthcare system capable of rational, bureaucratically accountable production on the one hand and real system production capabilities on the other creates interesting effects on the work floor. And these effects actually support the idea of autonomous, uniquely endowed clinical competence. For clinicians, competence is in part the ability to create successful clinical outcomes *despite* their organization—despite their managers and their computerized systems and processes, despite uninformed administrators and callous hospitals. McDonald and her colleagues (R. McDonald et al., 2006) observed a urology procedure in which the surgeon did not have access to the necessary equipment. The surgeon was not fazed, instead modifying an existing but nonstandard device in an attempt to continue with the operation. "Necessity is the mother of invention," he commented (p. 186), "that is what surgery is all about." Such extemporization to overcome odds and systemic constraints is closely tied to the identity of the physician as competent and unique:

> Accounts given by doctors convey the impression that they are competent professionals who struggle to achieve success in the context of factors which they are powerless to influence. Inadequate equipment, excessive workloads, inexperienced support staff, patients arriving late or having unpredictable anatomies are all factors over which doctors appear to have no control. (p. 197)

One way of looking at this is that organizational constraints, political expectations, and the bureaucratic accountability of the system put factors beyond the reach of medical competence. A clinician can try to focus all he or she wants and be as diligent as possible, but it is not going to change the fact that there are 20 more patients waiting outside, that a physician has not learned how to respond to an assertive remark about latex allergy, that the index of the drug is so low that the therapeutic dose is close to the toxic dose, or that there simply is no bed available in postoperative recovery. The lack of control over these factors implies that competence is only seldom the sole factor that mediates between successful or unsuccessful outcomes.

But the narrative of medical competence turns these constraints into a confirmation of itself. These factors just form the backdrop against which (or despite which) good performance must be achieved. The constraints of the system do not put things beyond the control and competence of medical practitioners but actually force them to evolve and fine-tune *additional* layers of creative competence. Competence at extemporizing, adapting, improvising, making do, and still succeeding. This construction preserves the image of the supremacy of medical competence and its special status within a system that does not understand the messy details of medical practice.

This is in a sense the heroic narrative of competence that also animates other safety-critical professions, particularly those that tend to employ highly autonomous, independent, controlling individuals. Air traffic controllers, for example, consistently report that the major source of their work-related stress is not peak traffic, but the organization, its management, staff shortages, and the lousy system interfaces with which they have to work (Stokes & Kite, 1994). Professional identity and peer reputation are closely linked to the evolved and fine-tuned ability to "push tin" (move air traffic along) despite these obstacles.

In nursing, the "hiding" of critical and scarce resources such as recovery beds is another example of getting the system to work by not following formal protocol. Recovery beds, for example, can be marked up with a status that makes them unusable for normal situations (e.g., cleaning) but instead preserves the resources so that they can absorb the unpredictable ebb and flow of postoperative patients. Such actions prevent the system from getting into gridlock by inserting an invaluable layer of redundancy and resilience. Managers would typically see resource wastage and might do anything to try to control the beds (Cook & Rasmussen, 2005).

Aircraft maintenance is another area in which succeeding despite the organization and despite the rules is a source of professional pride and interpeer comparison. In that world, a so-called job perception gap exists in which supervisors are convinced that safety and success result from mechanics following procedures—a sign-off means that applicable procedures were followed. But mechanics may encounter problems for which the right tools or parts are not at hand; the aircraft may be parked far away from base. Or there may be too little time: Aircraft with a considerable number of problems may have to be turned around for the next flight, and delays are discouraged. Mechanics consequently see success as the result of their evolved skills at adapting, inventing, compromising, and improvising in the face of local pressures and challenges on the line. A sign-off means the job was accomplished in spite of resource limitations, organizational dilemmas, and pressures. Those mechanics who are most adept are valued for their productive capacity even by higher organizational levels. Unacknowledged by those levels, though, are the vast informal work systems that develop so mechanics can get work done. Mechanics advance their skills at improvising, satisfying while sacrificing, and impart them to one another and condense them in unofficial, self-made documentation (N. McDonald, Corrigan, & Ward, 2002).

Informal work systems emerge and thrive because written guidance is inadequate to cope with local challenges and surprises and because the conception of work according to written guidance collides with the scarcity, pressure, and multiple goals of real work. Seen from the outside, a defining characteristic of informal work systems is their routine nonconformity.

Managers may be quick to point out that the system is actually more linear and predictable than many clinicians think. From the managers' or administrators' point of view, if only clinicians would follow the rules and play as a team, then management would attain the control needed to make the system better for everybody. Note how this is in part a depiction of powerlessness by the managers, who thereby insulate themselves from any critique when things fail (R. McDonald et al., 2006). From the clinician's point of view, the same "nonconformity" is a mark of expertise, fueled by professional and interpeer pride. This is consistent with the image of the physician as artisan, as independent and improvising hero (Gawande, 2002; R. McDonald & Harrison, 2004).

This does not mean that safety and economics are only marginal concerns for those who practice medicine. On the contrary, it could be argued that both safety and economic pressures are central to a medical worldview. Medical work is inherently economic and inherently about safety. It can be about implicitly or explicitly trading off interventions and clinical pathways against chances of success vis-à-vis a patient's condition, age, opportunities for recovery, and so forth. Medicine is problem solving under constraints and uncertainty, economic and otherwise. This may give rise to heroic narratives and constructions of identity, but it also remains a consistent feature in how practitioners see their daily work and negotiate the various actions they can or will take.

Problem solving under constraints means creating treatment plans or other interventions to minimize the chances of failure (economical, clinical) and devising them in a way that is both safe and economically responsible. Of course, these requirements conflict and cannot be reconciled. To paraphrase Petroski (1985), all clinical interventions are in some degree failures because they flout one or another of the requirements of practicing perfection. They are always compromises, and compromises imply some degree of failure.

Capitalistic concerns about costs and efficiencies are not alien to those who practice medicine. Respect for, and contribution to, rules, deadlines, project assignments, and the management of resource constraints is something that emerges from the practice of medicine as well. These are not just inconveniences that are imposed from the top down but are seen as inevitable and legitimate concerns. Medical practitioners will often accept the working conditions created by upper echelons, including production pressures, cost cutting, overtime, limited resources, and other kinds of compromises. In fact, support for hierarchical arrangements that attempt to deal with those things can even grow out of practitioners' aspirations of upward mobility inside the administration of a hospital. Hierarchies are structures not only of command and control but also of opportunity.

Given these reflections, the human factors approach to patient safety is not about a simple-minded or traditional imposition of rules for how to practice, as Berwick (2003) pointed out years ago. Indeed, human factors has become nuanced in its understanding of the relationship between written guidance and actual work (see Chapters 5 and 6). For instance, it no longer sees the development of protocol or guideline as the application of a mechanistic, static view of one best practice from the top down. In fact, there is always a distance between written guidance and actual work, and the application of guidelines, checklists, or procedures is a substantive cognitive activity that takes coordinative and interpretive work (Dekker, 2003).

STANDARDIZATION: YES OR NO?

Although Table 1.1 consistently lists "no" under the medicine column, moves also are accelerating to introduce competency checking, teamwork, and communication training in healthcare. Simulator centers are sprouting up across the world; practitioners can train and hone skills in situations that carry no jeopardy for a patient (Carroll & Messenger, 2008; Wagner, Hallmark, Farrar, & Overstreet, 2008; Wang et al., 2008). Systematic training of technical skills is also being addressed in fields such as surgery (Aggarwal, Grantcharov, & Darzi, 2007), as are those related to emergencies that arise from equipment failure in anesthesia (Waldrop, Murray, Boulet, & Kras, 2009). There is some modest but growing use of checklists (Haynes et al., 2009).

There are limits to the usefulness of standardization in any complex, dynamic field of activity. The premise that animates standardization, after all, is that variety and diversity are bad, and that deviance is unsafe. This is not necessarily true, and enforcing that belief too strictly can eventually erode the ability of a system to explore, adapt, and learn. Also, the initial enthusiasm with standardization in healthcare may have been based on the assumption that errors and deviations are the result of limitations in the cognitive psychological or social abilities of the practitioners working there. Limiting the capacity for individual discretion and diversion is thought to be the answer (R. McDonald et al., 2006). This, as the latest developments in human factors show, is not necessarily true (see Chapter 8).

Nevertheless, there are tasks and situations for which *not* having standardized responses (even as a basis for deviations) can easily be judged as unprofessional or even ethically problematic. The standardization of trauma life support was controversial when it was first proposed. Applying normal but diverse diagnostic approaches that assembled all relevant tests and patient data before devising a treatment plan was slowly but surely edged out. A standardized approach that focused the attention of the trauma team on the succession of main killers has taken over: airway, breathing, circulation, disabilities, exposure. There are now few first responders or emergency department workers who do not know their ABCDEs.

There are also widespread variations in some practices or procedures across hospitals or even towns or regions that really do not make any clinical sense. Other variations can be ascribed to ethnic or racial bias (Wennberg, Freeman, & Culp, 1987).

In both cases, moves to standardize practice would seem justified. This would also hold, for example, for the use of a surgical checklist with persuasive evidence that it saves lives (Gawande, 2010; Haynes et al., 2009). It can also hold for the use of some standard heuristics in clinical handovers that have been shown to work (Patterson, Roth, Woods, Chow, & Gomes, 2004). At a minimum, this can include avoiding interruptions and simply building in a process for verification and an opportunity to ask and answer questions (Wachter, 2008). It also goes for the standardized use of names and places for people, equipment, and tools that are used by an emergency response team or for the use of protocols and checklists in the blood bank at a hospital.

After all, not everything about a complex, dynamic field of activity is complex or dynamic. There are islands of stability and repetition for which standardized and simplified responses make sense. They can provide a layer of redundancy and double checking: a resilient foundation for confronting clinical diversity and adversity. These responses are tried, tested, and available—in a well-developed way. So, why are they not implemented? Amalberti and his colleagues (2005) explored the systemic barriers to achieving ultrasafe healthcare. They also showed that a lack of progress on safety in healthcare in general (and its wide diversity among specialties in particular) is linked not so much to an unavailability of pertinent tool kits (although individual practitioners may have no idea about their existence). Rather, it is linked to cultural and historical precedents and beliefs about performance and autonomy in medicine, which in turn are tightly interwoven with physician identity.

THE EXPECTATION OF PERFECTION VERSUS THE INEVITABILITY OF MISTAKE

It has been said that infallibility is the working hypothesis of the medical profession. Lucien Leape, in his 1994 reflections on human error in medicine, considered how

> the most important reason physicians and nurses have not developed more effective methods of error prevention is that they have a great deal of difficulty in dealing with human error when it does occur. The reasons are to be found in the culture of medical practice. Physicians are socialized in medical school and residency to strive for error-free practice. There is a powerful emphasis on perfection, both in diagnosis and treatment. In everyday hospital practice, the message is equally clear: mistakes are unacceptable. Physicians are expected to function without error, an expectation that physicians translate into the need to be infallible. One result is that physicians, not unlike test pilots, come to view an error as a failure of character—you weren't careful enough, you didn't try hard enough. This kind of thinking lies behind a common reaction by physicians: "How can there be an error without negligence?" (p. 1851)

There is nothing wrong with striving for perfection, of course. It is consistent with the idea of medicine as a vocation, a calling, rather than normal technical work. Assertions of infallibility also have been made on behalf of other status-determined professions in the past, including that of the pope. As Charles Vincent (2006) pointed out for medicine:

> Those working in this environment foster a culture of perfection, in which errors are not tolerated, in which a strong sense of personal responsibility both for errors and outcome is expected. ... With this background it is not surprising that mistakes are hard to deal with, particularly when so much else is at stake in terms of human suffering. (p. 142)

Failures, then, can get to be regarded as nonexistent anomalies in the system. This makes them, in principle, not reportable; they cannot be talked about for what they are. The existence of a "hidden curriculum" (Karnieli-Miller, Vu, Holtman, Clyman, & Inui, 2010), which teaches medical students and residents a repertoire of actions and vocabulary of phrases to deal with the inevitable imperfections of medical practice, is another example. The very idea that this curriculum has to be hidden and discourse about failure wrapped in euphemisms that obfuscate practitioner agency (clinical outcomes are not as expected because of "complications" or a "nonconforming patient") is testimony to the difficulty of integrating the notion of imperfection as a fundamental part and limitation of medical work.

But there is a contradiction here. Almost everybody in healthcare will readily acknowledge that "nobody is perfect, you will make an error occasionally although you take all precautions to avoid it" (R. McDonald et al., 2006, p. 192). In other words, mistake is inevitable, but the practice of medicine should aim for perfection, for infallibility. How does infallibility as a working hypothesis get reconciled with the inevitability of mistake? There are least two ways, neither of which encourage the reporting of errors and near misses.

The first way is by appealing to incompetence—either remediable clinical inexperience or irredeemable unsuitability for the profession (Bosk, 2003). In the first case, the person involved in a near miss is considered not good enough (yet) and still has a lot to do to achieve levels of perfection. It causes a senior practitioner to counsel, help, direct, or take over altogether. There is the expectation that the frequency of such events decreases as time passes and as the individual's experience accumulates. Again, infallibility could be preserved as a working hypothesis. Achieving it is just pushed into the future a bit.

Technical Errors and Need for More Instruction to Achieve Perfection

Bosk (2003) told how Carl, a surgical intern, was closing an incision, while Mark, the chief resident, was assisting. Carl was ill at ease. He turned to Mark and said, "I can't do it." Mark said, "What do you mean, you can't? Don't ever say you can't. Of course you can." "No, I just can't seem to get it right." Carl had been forced to put in and remove stitches a number of times, unable to draw the skin closed with the proper tension. Mark replied, "Really, there is nothing to it," and taking Carl's hand in his own, he said, "The trick is to keep the needle at this angle and put the stitch through like this," all the while leading Carl through the task. "Now, go on." Mark then let Carl struggle through the rest of the closure on his own.

In another example, Bosk told of the difficulty of performing a myelogram (a diagnostic procedure involving the removal of spinal fluid and the injection of dye in the spinal column) that had been ordered for a patient named Mr. Eckhardt. A senior student was to instruct a junior student in the procedure. They tried without any success to get the needle in the proper space. After some fumbling and a few sticks at the patient, the senior student instructed the junior student to go "get Paul" (a second-year resident). Paul came in and surveyed the situation. After examining Mr. Eckhardt's back, he told the students, who were profusely apologizing for their failure, not to worry; the problem was in Mr. Eckhardt's anatomy and not in their skills. He then proceeded with some difficulty to complete the procedure, instructing the students all the time (Bosk, 2003, pp. 44–45).

If the individual did not discharge role obligations diligently and created unnecessary extra work for colleagues, then the performance would be considered normatively erroneous. Bosk found how senior surgeons continuously made assessments about whether the person making the mistake actually has any business being in the profession in the first place. Mistakes were implicitly classified in these assessments, with different repertoires of action appended to them. The more such normative errors a person was seen to make, the less likely it would be that the person would retain employment within that surgical service (Bosk, 2003). Infallibility also could be preserved here because it was a matter of particular individuals not belonging in the profession or in that particular specialty or service (the person may have lacked the "calling").

Normative Errors: Not Living Up to the Role of Clinician

One of the important normative aspects of learning surgery is to be honest about mistakes, but this goes to the heart of the paradox. Errors are impossible because of the infallibility working hypothesis. But they are inevitable. This can put junior doctors in interesting situations. "Covering up is never really excusable," Bosk (2003) quoted an attending physician as saying. The attending continued:

> You have to remember that each time a resident hides information, he is affecting someone's life. Now in this business it takes a lot of self-confidence, a lot of maturity, to admit errors. But that's not the issue. No mistakes are minor. All have a mortality and a morbidity. Say I have a patient who comes back from the operating room and he doesn't urinate. And say my intern doesn't notice or he decides it's nothing serious and he doesn't catheterize the guy and he doesn't tell me. Well, this guy's bladder fills up. There's a foreign body and foreign bodies can cause infections; infection can become sepsis; sepsis can cause death. So the intern's mistake here can cause this guy hundreds of dollars in extra hospitalization and it could cost him his life. All mistakes have costs attached to them. Now a certain amount is inevitable. But it is the obligation of everyone involved in patient care to minimize mistakes. (pp. 60–61).

To the extent that errors are the result of inexperience and lack of exposure, the solution is simply more training and more exposure. But if near misses are judged to result from the individual's fundamental unsuitability for the profession, then there is nothing systemic or deeper about the error that is interesting to probe.

The source of failure lies with the individual, and the solution is not to have the individual practice in that specialty or service (Bosk, 2003).

The other way in which the inevitability of mistake can be reconciled with infallibility as working hypothesis is to blame bad luck—arising from a particular set of circumstances: "People don't start off to harm a patient, it just happens" (R. McDonald et al., 2006, p. 192). In this case, the near miss can be said to be something external, something accidentally imported through happenstance or bad luck. If a near miss is ascribed to a highly unusual set of circumstances, then there is not much reason to reflect, probe, and learn. The rationalization is that this particular set of circumstances (anatomical, physiological, operational, team composition, organizational, and so forth) will not likely repeat itself in exactly this way, so there is little value in trying to predict and prevent it from creating trouble again. Indeed, it also is not useful to share any lessons from that particular encounter with those circumstances with colleagues because they will not likely meet them, and in the end errors are inevitable anyway:

> If mistakes are seen as inevitable and a matter of bad luck arising from a particular set of circumstances, then this implies that attempting to learn from other people's mistakes (and by implication, reporting those mistakes) is not regarded as a valuable exercise. (R. McDonald et al., 2006, p. 194)

These various repertoires of reconciliation have a consequence for the ability of the medical profession to seriously reflect on mistakes and learn from them. Critical reflection on safety and error is not a taken-for-granted feature of the professional identity of people working in healthcare. This makes it hard to legitimate the reporting and discussion of error (Berlinger, 2005). Even 2,000 years ago, the Roman writer Celsus had to encourage the honest disclosure of mistake. This was not for the benefit of the patient. Four centuries earlier, the Greek medical class also had not worried about that too much. Rather, such disclosure was for the benefit of peers, so that everybody could learn from everybody else's mistakes, and the whole profession would be wiser for it (Miles, 2004):

> Sincere confession of the truth befits a great mind which will still be ready to accept many responsibilities, and especially in performing the task of handing down knowledge for the advantage of posterity that no one else may be deceived again by what has deceived him. (p. 113)

Confessing error also was not uncommon in classical Greek medical writings. Errors were considered the inevitable byproduct of a complex and uncertain activity. Judgments of the severity of a head wound, or of surgical indications of a swollen lower abdomen, could as easily be wrong as they could be right (Miles, 2004). As affirmed by Celsus, such confession, at least in writing, was seen as noble. It was consistent with the acceptance of "many responsibilities," an obligation to colleagues to warn them against falling for these deceptions.

INFALLIBILITY, ERROR, AND TECHNICAL ADVANCES IN MEDICINE

Infallibility as the working hypothesis of medicine has gone through a long evolution. The expectation of perfection is not static: It has evolved on the back of technical and knowledge advances in the practice of medicine. Healers in premodern societies may have claimed infallibility, for sure, even if their interventions were unsuccessful. An appeal to divine or diabolic intervention could always override the healer's individual shortcomings. Also in the West, people have long explained disease and misfortune through divine or diabolic intervention in their lives (Green, 2003). It was not until the nineteenth century that the grip of theodicy started loosening in the West, which gradually replaced its inevitable fatalism with a more optimistic view of human control and progress (DeVille, 2004).

Even with the growth of modern medicine, perfection actually was not initially expected. People understood the limitations of clinical knowledge and skills and were aware of the lack of a coherent set of medical explanations or procedures for dealing with disease. Body cavity surgery, for example, was hardly practiced before the 1880s. Death from the effects of peritonitis or intestinal blockage was not seen as an error made by medicine or any of its practitioners but rather the result of a poorly understood medical condition for which no cure was easily available. Even when surgical intervention was attempted, its likely failure was not ascribed to medical error, and malpractice suits were uncommon (DeVille, 2004).

But as medical knowledge and technical capabilities kept expanding, they gradually colonized and clarified previously nebulous or inaccessible areas of human anatomy and physiology and their illnesses. From the 1920s, surgical intervention became more common and more successful. Sulfa drugs, aseptic practices, transfusions, better training, and better instruments all helped surgeons achieve better operative outcomes (DeVille, 2004).

With expanding techniques and knowledge came two things: The first was a vastly expanded repertoire of skills, complex procedures, knowledge demands, and technical tools. These were not only the sources of surgical success but also new (and many more) areas in which things could go wrong. They allowed surgeons new opportunities not only to express expertise but also to commit error. The second thing that came with medical advances was the increase in patient expectations and demands. Diseases that were previously beyond the realm of intervention, and equal to a death sentence, were now seen to be under (a modicum of) medical control. If outcomes still were not good, and the expectations were violated, it would be more likely for patients to seek the source of such failures in medical errors rather than in the complexity or diversity of the disease.

Orthopedic Advances

The recent history of orthopedics illustrates how expanding knowledge and skills are a double-edged sword when it comes to medical error (DeVille, 2004). At the end of the eighteenth century, the standard procedure for dealing with serious fractures or dislocations was amputation. This brought disfigurement, disability, and often death, but the procedure resulted in basically no malpractice suits.

Patient expectations were as low as the skills of the physicians from whom they sought help.

By the late 1830s, however, the treatment of such fractures and dislocations became to be seen as a relatively simple mechanical procedure that in many cases obviated the need for amputation of the limb. In fact, patients could often expect perfect cures.

The new procedures did create new knowledge and skill and competence demands on the part of orthopedic surgeons, confronting them with a new range of possible complications and complexities during repair, setting, and convalescence. Patient expectations went up. Side effects (unusually long convalescence, frozen limbs) were no longer seen as the natural byproduct of an uncertain medical intervention but as evidence of a lack of skill or competence on the part of the medical practitioner who did the intervention. Malpractice suits resulting from orthopedic injuries were the most common type of claims through the late 1930s.

Other advances, particularly in diagnostics during the latter part of the twentieth century, have put the effects (good or bad) of medical interventions on full display for patient and clinician alike. Ultrasound, bronchoscopy, endoscopy, magnetic resonance imaging, and computed tomography all showed what was done and how well it was done, leaving a much clearer trace of evidence for anybody wanting to assert "medical error." Medical advances, in other words, created greater opportunities for both expertise and error, enhanced their visibility, and inflated the expectations of success on the part of the patient. This was a potent combination that accelerated the image of a medical error problem. It was accompanied by an increase in malpractice suits and more recently, the trend toward criminalization of medical error (Dekker, 2007; DeVille, 2004).

The cycle does not have to stop there, however. For instance, it did not for laparoscopic cholecystectomy, a procedure for the removal of the gallbladder that was introduced in the late 1980s. It was a much less invasive technique than open surgery and created great patient expectations of no complications and quick recoveries. Not long after the introduction, however, it became obvious that the laparoscopic procedure entailed entirely new risks, particularly in anatomic identification (Hugh & Dekker, 2008). Surgically induced bile duct injuries, for example, appeared to be a particular byproduct of using a laparoscope (although the data are not convincing on this). As a result, claims for incompetence and malpractice soared by 500% as compared to the previous open-surgery cholecystectomy procedure (DeVille, 2004). Years after the introduction of the procedure, however, knowledge and discourse on how to prevent such injuries had increased, and awareness of the risks had grown on the part of surgeons, who in turn could also better inform their patients about the drawbacks and benefits of the procedure (Hugh, 2002). With this additional growth in knowledge, demands and expectations of perfection were modulated downward.

This is likely an ingredient in a much larger shift in how medical competence is viewed today. During the last half of the twentieth century, Western societies increasingly began to see the provision of their medical care as based on a contract rather than on the status of the provider. The conception of care—how to get it, where to get it, from whom to get it—has increasingly come to be seen as a commercial trans-action between citizens, a contract, whether explicit or not, in which promise and delivery are concretized and matched. The explosion of Internet medicine, in which patients are in certain cases considerably better informed about their own conditions than their doctors, has contributed powerfully to this leveling and democratization of the relationship between patient and caregiver. This transformation is gradually devaluing the special role and the unique, sacred status of the healer in Western soci-ety. It is an invaluable contribution to our ability to make progress on patient safety. No longer do we need to see expertise and error in healthcare as the celebration or devastating violation of a sacred calling. Rather, we can begin to understand how expertise and error in healthcare, like in any safety-critical endeavor, are normal, lawful expressions of highly skilled, technical work. We can begin to learn how both error and expertise are systematically connected to features of the tools, tasks, and operating environment in which we expect people to carry out their work.

KEY POINTS

- In few other safety-critical fields is the simultaneous belief in individual strength and brittleness as strong as in healthcare. People in healthcare often think that safety lies foremost in the hands through which care ulti-mately flows to the patient. That is the point at which we should intervene to make things safer, to tighten practice, and to focus attention.
- Medicine has retained a unique subculture with its own rules, norms, mythologies, social structures, hierarchies, clothing and tools, and other markers of status and specialization and identity. Physicians have generally resisted the notion that their relationship with patients is anything like a commercial or contractual one, relying rather on what they and society see as status-determined duties, sometimes invoking a "sacred calling."
- Healthcare is special in its institutional and cultural assumptions about competence. For example, once a clinician has learned something, there is little agreement on a need to periodically rehearse or check that skill extensively. Also, not many questions might be asked whether the clinician is competent to tackle a task or technology never done or used before.
- Patients are, in a gross characterization, exposed to three kinds of risk—their disease, the diagnosis and treatment plan for it, and the implemen-tation of that plan. There is a persistent idea that, to handle these risks, physicians have to be unique craftspeople whose exercise of skill is about situational insight, deftness, contextual sensitivity, mastery, and prowess. As a result, healthcare has a unique and complicated relationship to issues of standardization and competence. There is a strong preoccupation with the autonomy and discretion of its individual actors.
- When things go well, healthcare tends to celebrate "good doctoring"—acts by competent people who succeeded despite the organization and its com-plexity. When things do not go well—when adverse events occur—health-care tends to zero in on the people at the sharp end who, for once, failed

to hold that complex, pressurized patchwork together. The human factors approach, instead, wants to question and investigate the systemic sources of all that complexity.

REFERENCES

Aggarwal, R., Grantcharov, T. P., & Darzi, A. (2007). Framework for systematic training and assessment of technical skills. *Journal of the American College of Surgeons, 204*(4), 697–705.

Amalberti, R. (2001). The paradoxes of almost totally safe transportation systems. *Safety Science, 37*(2–3), 109–126.

Amalberti, R., Auroy, Y., Berwick, D., & Barach, P. (2005). Five system barriers to achieving ultrasafe healthcare. *Annals of Internal Medicine, 142*(9), 756–764.

Arvigo, R., Epstein, N., & Yaquinto, M. (1994). *Sastun: My apprenticeship with a Maya healer.* San Francisco: HarperSanFrancisco.

Berlinger, N. (2005). *After harm: Medical error and the ethics of forgiveness.* Baltimore: Johns Hopkins University Press.

Berwick, D. M. (2003). Improvement, trust, and the healthcare workforce. *Quality and Safety in Health Care, 12*(6), 448–452.

Bosk, C. (2003). *Forgive and remember: Managing medical failure.* Chicago: University of Chicago Press.

Bradley Hagerty, B. (Writer). (2010, April 12). Vatican: Bishops must report alleged abuse to police [Radio broadcast]. Washington, DC: NPR.

Cacciabue, P. C., & Vella, G. (2010). Human factors engineering in healthcare systems: The problem of human error and accident management. *International Journal of Medical Informatics, 79*(4), 1–17.

Campbell, R. D., & Bagshaw, M. (1991). *Human performance and limitations in aviation.* Oxford, UK: Blackwell Science.

Carroll, J. D., & Messenger, J. C. (2008). Medical simulation: The new tool for training and skill assessment. *Perspectives in Biology and Medicine, 51*(1), 47–60.

Cohen-Charash, Y., & Spector, P. E. (2001). The role of justice in organizations: A meta-analysis. *Organizational Behavior and Human Decision Processes, 86*(2), 278–321.

Conley, R. J. (2005). *Cherokee medicine man: The life and work of a modern-day healer.* Norman: University of Oklahoma Press.

Cook, R., & Rasmussen, J. (2005). "Going solid": A model of system dynamics and consequences for patient safety. *Quality and Safety in Health Care, 14*(2), 130–134.

Davis-Floyd, R., & St. John, G. (1998). *From doctor to healer: The transformative journey.* New Brunswick, NJ: Rutgers University Press.

Dekker, S. W. A. (2003). Failure to adapt or adaptations that fail: Contrasting models on procedures and safety. *Applied Ergonomics, 34*(3), 233–238.

Dekker, S. W. A. (2007). Criminalization of medical error: Who draws the line? *ANZ Journal of Surgery, 77*(10), 831–837.

DeVille, K. (2004). God, science, and history: The cultural origins of medical error. In V. A. Sharpe (Ed.), *Accountability: Patient safety and policy reform* (pp. 143–158). Washington, DC: Georgetown University Press.

Drazen, J. M., & Epstein, A. M. (2002). Rethinking medical training: The critical work ahead. *New England Journal of Medicine, 347*, 1271–1272.

Dunn, E., & Wolfe, J. (2002). What's not wrong with this prescription? *Drug Topics, 146*(9), 25–26.

Farrington-Darby, T., & Wilson, J. R. (2006). The nature of expertise: A review. *Applied Ergonomics, 37*, 17–32.

Federal Aviation Administration. (1996). *Federal Aviation Administration Human Factors Team report on the interfaces between flightcrews and modern flight deck systems.* Washington, DC: Author.

Fonarow, G. C., Abraham, W. T., Albert, N. M., Stough, W. G., Gheorghiade, M., Greenberg, B. H., et al. (2007). Association between performance measures and clinical outcomes for patients hospitalized with heart failure. *Journal of the American Medical Association, 297*(1), 61–70.

Gander, P. H., Merry, A., Millar, M. M., & Weller, J. (2000). Hours of work and fatigue-related error: A survey of New Zealand anaesthetists. *Anaesthesia and Intensive Care, 28*(2), 178–183.

Gawande, A. (2002). *Complications: A surgeon's notes on an imperfect science.* New York: Picador.

Gawande, A. (2008). *Better: A surgeon's notes on performance.* New York: Picador.

Gawande, A. (2010). *The checklist manifesto: How to get things right.* New York: Metropolitan Books.

Green, J. (2003). The ultimate challenge for risk technologies: Controlling the accidental. In J. Summerton & B. Berner (Eds.), *Constructing risk and safety in technological practice.* London: Routledge.

Haynes, A. B., Weiser, T. G., Berry, W. R., Lipsitz, S. R., Breizat, A. H., Dellinger, E. P., et al. (2009). A surgical safety checklist to reduce morbidity and mortality in a global population. *New England Journal of Medicine, 360*(5), 491–499.

Helmreich, R. L. (2000). On error management: Lessons from aviation. *British Medical Journal, 320*(7237), 781–785.

Hirshhorn, C. (2000). Poor penmanship costs MD $225,000. *CMAJ: Canadian Medical Association Journal, 162*(1), 91–92.

Hugh, T. B. (2002). New strategies to prevent laparoscopic bile duct injury—Surgeons can learn from pilots. *Surgery, 132*(5), 826–835.

Hugh, T. B., & Dekker, S. W. A. (2008). Laparoscopic bile duct injury: Understanding the psychology and heuristics of the error. *ANZ Journal of Surgery, 78*(12), 1109–1114.

Karnieli-Miller, O., Vu, T. R., Holtman, M. C., Clyman, S. G., & Inui, T. S. (2010). Medical students' professionalism narratives: A window on the informal and hidden curriculum. *Academic Medicine, 85*(1), 124–133.

Klein, G. A. (1993). A recognition-primed decision (RPD) model of rapid decision making. In G. A. Klein, J. Orasanu, R. Calderwood, & C. E. Zsambok (Eds.), *Decision making in action: Models and methods* (pp. 138–147). Norwood, NJ: Ablex.

Kohn, L. T., Corrigan, J., & Donaldson, M. S. (2000). *To err is human: Building a safer health system.* Washington, DC: National Academy Press.

Landrigan, C. P., Rothschild, J. M., Cronin, J. W., Kaushal, R., Burdick, E., Katz, J. T., et al. (2004). Effect of reducing interns' work hours on serious medical errors in intensive care units. *New England Journal of Medicine, 351*(18), 1838–1848.

Leape, L. L. (1994). Error in medicine. *Journal of the American Medical Association, 272*(23), 1851–1857.

McDonald, N., Corrigan, S., & Ward, M. (2002). *Well-intentioned people in dysfunctional systems.* Paper presented at the Fifth Workshop on Human Error, Safety and Systems Development, June 2002, Newcastle, Australia.

McDonald, R., & Harrison, S. (2004). The micropolitics of clinical guidelines: An empirical study. *Policy and Politics, 32*(2), 223–229.

McDonald, R., Waring, J., & Harrison, S. (2006). Rules, safety and the narrativization of identity: A hospital operating theatre case study. *Sociology of Health and Illness, 28*(2), 178–202.

Miles, S. H. (2004). *The Hippocratic oath and the ethics of medicine.* Oxford, UK: Oxford University Press.

Mulcahy, J. B. (2001). *Birth and rebirth on an Alaskan island: The life of an Alutiiq healer.* Athens: University of Georgia Press.

Nemeth, C., Nunnally, M., O'Connor, M., Klock, P. A., & Cook, R. (2005). Getting to the point: Developing IT for the sharp end of healthcare. *Journal of Biomedical Informatics, 38*(1), 18–25.

Outram, D. (2005). *The enlightenment* (2nd ed.). Cambridge, UK: Cambridge University Press.

Parker, D., & Lawton, R. (2000). Judging the use of clinical protocols by fellow professionals. *Social Science & Medicine, 51*(5), 669–677.

Patterson, E. S., Roth, E. M., Woods, D. D., Chow, R., & Gomes, J. O. (2004). Handoff strategies in settings with high consequences for failure: Lessons for health care operations. *International Journal for Quality in Health Care: Journal of the International Society for Quality in Health Care, 16*(2), 125–132.

Pellegrino, E. D. (2004). Prevention of medical error: Where professional and organizational ethics meet. In V. A. Sharpe (Ed.), *Accountability: Patient safety and policy reform* (pp. 83–98). Washington, DC: Georgetown University Press.

Pellegrino, E. D., Thomasma, D. C., & Kissell, J. L. (2000). *The health care professional as friend and healer: Building on the work of Edmund D. Pellegrino.* Washington, DC: Georgetown University Press.

Peterson, G. M., Wu, M. S. H., & Bergin, J. K. (1999). Pharmacists' attitudes towards dispensing errors: Their causes and prevention. *Journal of Clinical Pharmacy and Therapeutics, 24*(1), 57–71.

Petroski, H. (1985). *To engineer is human: The role of failure in successful design.* New York: St. Martin's Press.

Pilz, A. (1988). *Manang Jabing Anak Incham: A study of an Iban healer/Sarawak.* Berlin: Reimer.

Pink, D. H. (2005). *A whole new mind: Moving from the information age to the conceptual age.* New York: Riverhead Books.

Pronovost, P. J., & Vohr, E. (2010). *Safe patients, smart hospitals.* New York: Hudson Street Press.

Rawls, J. (2003). *A theory of justice.* Cambridge, MA: Harvard University Press.

Rochlin, G. I., LaPorte, T. R., & Roberts, K. H. (1987). The self-designing high reliability organization: Aircraft carrier flight operations at sea. *Naval War College Review*, 76–90.

Rogers, A. E., Hwang, W. T., Scott, L. D., Aiken, L. H., & Dinges, D. F. (2004). The working hours of hospital staff nurses and patient safety. *Health Affairs (Project Hope), 23*(4), 202–212.

Stewart, P. J., & Strathern, A. (2004). *Witchcraft, sorcery, rumors, and gossip.* Cambridge, UK: Cambridge University Press.

Stokes, A., & Kite, K. (1994). *Flight stress: Stress, fatigue, and performance in aviation.* Aldershot, UK: Avebury Aviation.

Sugden, C., Aggarwal, R., & Darzi, A. (2010). Re: Sleep deprivation, fatigue, medical error and patient safety. *American Journal of Surgery, 199*(3), 433–434.

Tammelin, A., Karell, A.-C., Nilsson, P., Samuelson, A., Tillman, E., & Örtqvist, Å. (2007). Sjukvårdsklädsel måste förena bra förtroende och god hygien [Doctor's dress must combine confidence with good hygiene]. *Läkartidningen, 104*(5), 350–351.

Vaughan, D. (1996). *The Challenger launch decision: Risky technology, culture, and deviance at NASA.* Chicago: University of Chicago Press.

Velo Giampaolo, P., & Minuz, P. (2009). Medication errors: Prescribing faults and prescription errors. *British Journal of Clinical Pharmacology, 67*(6), 624–628.

Vincent, C. (2006). *Patient safety.* London: Churchill Livingstone.

Vogel, L. (2010). Nursing degree "opens doors beyond bags, beds and bedpans." *CMAJ: Canadian Medical Association Journal, 182*(2), 131–132.

Wachter, R. M. (2008). *Understanding patient safety.* New York: McGraw-Hill.

Wagner, L. J., Hallmark, B., Farrar, C., & Overstreet, M. (2008). Tennessee Nursing Partnership promotes skill-advancement in simulation technology for nurse educators in Tennessee. *Tennessee Nurse/Tennessee Nurses Association, 71*(3), 1, 5.

Waldrop, W. B., Murray, D. J., Boulet, J. R., & Kras, J. F. (2009). Management of anesthesia equipment failure: A simulation-based resident skill assessment. *Anesthesia and Analgesia, 109*(2), 426–433.

Wang, E. E., Quinones, J., Fitch, M. T., Dooley-Hash, S., Griswold-Theodorson, S., Medzon, R., et al. (2008). Developing technical expertise in emergency medicine—The role of simulation in procedural skill acquisition. *Academic Emergency Medicine, 15*(11), 1046–1057.

Wears, R. L., & Perry, S. J. (2002). Human factors and ergonomics in the emergency department. *Annals of Emergency Medicine, 40*(2), 206–212.

Welch, D. L. (1998). Human factors usability test and evaluation. *Biomedical Instrumentation & Technology/Association for the Advancement of Medical Instrumentation, 32*(2), 183–187.

Wennberg, J. E., Freeman, J. L., & Culp, W. J. (1987). Are hospital services rationed in New Haven or over-utilised in Boston? *Lancet, 1*(8543), 1185–1189.

Wong, D., Nye, K., & Hollis, P. (1991). Microbial flora on doctors' white coats. *British Medical Journal, 303*(6817), 1602–1604.

Yentis, S. (2010). Of humans, factors, failings and fixations. *Anaesthesia, 65*(1), 1–4.

Young, D. E., Ingram, G., & Swartz, L. (1989). *Cry of the eagle: Encounters with a Cree healer.* Toronto: University of Toronto Press.

2 The Problem of "Human Error" in Healthcare

NUMBERS ARE STRONG

"Is it not really strange," Albert Einstein asked in a 1952 letter to fellow quantum physicist Max Born, "that human beings are normally deaf to the strongest of argument while they are always inclined to overestimate measuring accuracies?" The topic of Einstein's letter was a theoretical controversy about the bending of light near the sun, but his fascination about what gets people's attention has a general appeal. Numbers got people's attention for the patient safety problem whereas arguments had not. As early as the 1980s, a strong argument existed that much of the iatrogenic harm (i.e., harm caused by medical examination or treatment) is preventable (Leape et al., 1991). Yet it was not until almost a decade later that the results on which the argument were based, now known as the Harvard study, became more widely known. Its most celebrated quotation was not an argument, but a number: Between 44,000 and 98,000 people die each year because of preventable harm caused by medical care in the United States alone (Kohn, Corrigan, & Donaldson, 2000).

Replications across other developed nations quickly followed (Davis et al., 2002; Neale, Woloshynowych, & Vincent, 2001; Thomas, Studdert, Burstin, et al., 2000; Thomas, Studdert, Runciman, et al., 2000; Vincent, Neale, & Woloshynowych, 2001). Some were based on projection of the Harvard study results into the size of the populations and healthcare systems of other countries. The results were comparable across nations. The first national study in Canada on adverse events in acute care hospitals found that in 2000, of 2.5 million adult hospital admissions, 7.5% (185,000) resulted in an adverse event. Of these, 37% (70,000) were considered preventable. In total, the Canadians estimated that in one year between 9,000 and 24,000 of their patients experienced an adverse event that was preventable and later died (Baker et al., 2004). In England, one in ten patients was estimated to be hurt by medical care (Department of Health, 2000; Vincent et al., 2001).

Every Flight a Fatal Flight for Some

Put the numbers on iatrogenic harm in another context. After a flight from, say, New York to Miami, with an airplane that carried 150 passengers, only 148 emerge alive. Two have died simply because they were on the airplane. The flight alone caused a heart attack in one and turbulence-induced blunt-force head trauma in another. Four have developed infections because of being packed inside the hypoxic tube with bad air filtration and cabin crew who refused to wash their hands before serving snacks. Two of these infections are beyond the reach of antibiotics and will debilitate these people for life. One of these passengers has no choice: He will have to remain onboard the airplane for the rest of his life. Two passengers have been poisoned by badly mixed $7 cocktails, which caused permanent liver damage in one and stripped the stomach lining of the other. One has lost a leg from the hip down unnecessarily because it got trapped in the seat in front, another passenger had her common bile duct severed by a snagged seat belt, and yet another has suffered permanent brain damage because of oxygen supply problems near her seat. One child was electrocuted because of a short in the entertainment electronics circuit box mounted by her left ankle. Now, imagine the arrivals hall. What would the scene look like? Passengers are stumbling out in various states of disability and disease. Some are never going to come out. And this is not just one flight. It happens on every flight, every day, by every airline. Who would still fly?

Such numbers got people's attention. In the years following the publication of the Harvard study by the Institute of Medicine (Kohn et al., 2000), policy makers, politicians, the public, and the media slowly started waking up to the problem of patient safety. Iatrogenic harm took more lives than traffic accidents, breast cancer, or AIDS. The numbers represented not only unnecessary human suffering but also unnecessary economic loss. In England, adverse events were estimated to cost the National Health Service 2 billion pounds per year in additional hospitalization, even without considering the wider economic costs (Department of Health, 2000).

The numbers gave hospitals and administrations concrete goals to pursue. In 2004, Don Berwick, one of the pioneers in patient safety, put up the challenge to save 100,000 lives in the next 18 months (Berwick, Calkins, McCannon, & Hackbarth, 2006). Hospitals joining in the effort would have to deploy rapid response teams at the first sign of patient decline; deliver reliable evidence-based care for heart attacks; prevent medication errors, central line infections, surgical site infections, and ventilator-associated pneumonia—together the most powerful contributors to iatrogenic harm. Eighteen months later, Berwick announced that the results far exceeded expectations: Participating hospitals estimated that they had saved as many as 122,300 lives by implementing the recommended practices and strategies. The social pressure of an achievable, measurable, and comparable goal combined with an 18-month deadline put patient safety on the agenda of many hospital boards and administrations that might not have paid attention to it before (Berwick, Hackbarth, & McCannon, 2006).

Yet controversy about the numbers immediately erupted. The numbers were good for political and administrative action. They were good for mustering resources and managerial energy. But were they anywhere near accurate? How did participating hospitals measure a "saved" life? How exactly did they know when a life was saved from preventable harm? What was considered preventable by whom and when? Was this measured after the outcome or before? The number estimate could be a great exaggeration (Wachter & Pronovost, 2006).

Most studies on iatrogenic harm have kept the space between ceiling and floor rather generous. Put it all together, and we could say that between 3% and 16% of hospitalized patients are harmed by preventable medical errors. Why this range? The range is there because underneath the numbers runs a whole series of problematic issues related to assessing, classifying, categorizing, and counting instances of preventable harm and the putative medical errors that produce it. These issues call into question the very premise that there is such a thing as an "accurate" count of iatrogenic harm or preventable medical errors. Accuracy denotes precision and correctness (Golden-Biddle & Locke, 1993). Accuracy means that there is a "real" number of preventable medical errors out there in hospitals all around the country or the world—one definitive number. All we need to do is perfect the method to get at that number. The better the method, the more accurate the count. The better the method, the closer it takes us to that real number, to the absolute truth of iatrogenic harm.

Doctors Are 7,500 Times More Dangerous Than Gun Owners

There are about 700,000 physicians in the United States. The U.S. Institute of Medicine estimated that each year between 44,000 and 98,000 people die as a result of medical errors (Kohn et al., 2000). This makes for a yearly accidental death rate per doctor of between 0.063 and 0.14. In other words, up to one in seven doctors will kill a patient each year by mistake. In contrast, there are 80 million gun owners in the United States. They are responsible for 1,500 accidental gun deaths in a typical year (National Safety Council, 2004). This means that the accidental death rate caused by gun owner error is 0.000019 per gun owner per year. Only about 1 in 53,000 gun owners will kill somebody by mistake. Doctors, then, are 7,500 times more likely than gun owners to kill somebody as a result of human error.

Before an action or inaction gets assigned to the category "preventable error," a judgment needs to be made that indeed it *was* a preventable error. That judgment has to be made by someone. Was a central line infection in a patient preventable and the result of some erroneous placement procedure? Or, was it virtually inevitable—the normal result of putting a line in a patient with a poor prior condition on a busy ward? Do we get the answer from the attending, the resident, the nurse, the nurse manager, the lab tech who cultured the biopsy, the patient, the patient's husband? The question of who gets to say this muddles the possibility of one accurate count of iatrogenic harm or its causes. What are the background, the knowledge of patient and procedure, and the motives of the person making that judgment?

What is considered harm due to medical error, then, is almost always debatable. One reason is the contested nature of "error." Was an error made? Who says this? The nurse manager (see the opening to Chapter 1) asserted that the safety trouble on her ward was the result of nurses who make "mistakes." This inserts a ready-made judgment into any subsequent discussion. Characterizing actions or omissions as mistakes not only imbues them with a moral load they may not deserve but also basically excludes any other constructions of the same actions or omissions. Nurses themselves, for instance, might not see their work as consisting of a series of mistakes (some avoided, some not) but rather as a struggle to meet irreconcilable goals under enormous resource constraints and production pressures. Their construction of what is wrong would then give rise to a different discussion and set of countermeasures altogether.

Another reason is the contested relationship between the supposed error and a bad medical outcome. As shown in Chapter 1, there is a strong correlation between technical medical progress and the likelihood of interpreting outcome failures as medical error. Paget (2004) offered an even simpler reason for the contested relationship between error and outcome. The starting points for almost all healthcare are people who are already sick and its practice on human bodies. That means that bad outcomes are always possible as the normal, necessary by-product of the activity (Paget, 2004). Paget called medical work an:

> error-ridden activity because mistakes are indigenous to the work process. My characterization undermines the semantic sense of mistakes as uncommon, aberrant, or culpable acts. In saying this, I do not wish to imply that medical mistakes are never aberrant, culpable, or uncommon. Rather, it is the whole activity that is exceptional, uncommon, and strange because it is error-ridden, inexact, and uncertain, and because it is practiced on the human body. (p. 58)

THE HUMAN FACTORS APPROACH

To most people, though, human error is real and a real risk. Suppose that an anesthesia care provider has hooked up a patient to the anesthesia but had to twist the various cables and tubes to fit around the operating table; now, there is a kink in the tube that supplies oxygen. The patient's oxygen saturation quickly becomes problematic and even causes harm before the kink is discovered and corrected. It seems easy to deem the anesthetic hookup erroneous and to judge that the harm caused by it was preventable. Indeed, most people would say that an error was made: There was a mismatch between intention and outcome.

But there is a problem if we just stop at this point and attribute the iatrogenic harm to that error. Suppose the following: The arrangement of the table and the choreography of the surgical and anesthetic teams swarming around it also would have something to do with this situation. Then there was the decision by the manufacturer not to develop kink-resistant tubing or the decision by the management of the hospital not to purchase kink-free tubing once it became available. This occurred early on Sunday morning, following a 6-day week spent on a long surgical list, working through a backlog that resulted from resent political concern about waiting times, and after a particularly challenging operation that dragged out last night because of unexpected complications. There was also the decision not to train the surgical team in assertive communication, which meant that the anesthetist's attention was never meaningfully directed to the developing oxygen saturation problem when others saw the first signs of trouble. A preoperative time-out procedure was implemented not long ago but had proven ineffective as team members were never told what exactly they needed to review or remind each other of personally before the operation. The operating theaters in the hospital are located in a structure intended for other purposes. And of course, there are the issues of a computer interface that does not reveal oxygen saturation problems clearly and multiple warnings by the various anesthetic machines in the operating theater, all of them loud, underspecified, and intrusive warnings with a history of "crying wolf." An adverse event reporting system was in place in the hospital, but a nurse anesthetist who previously reported a kink in tubing had been reprimanded and stigmatized by the hospital's risk manager after intervention.

For some, all this would amount only to extenuating circumstances. These would be excuses. Excuses that should not have come in the way of doing the right thing and doing it in a timely manner. But since the 1940s an entire science has grown up around a manner of thinking to the contrary. This is human factors. Commenting on the advances in medicine during the latter half of the twentieth century, Alphonse Chapanis (2004, p. 11), one of the founding fathers of the field of human factors, explained:

> At about the same time as all these medical advances were being made, an entirely different technical discipline was born and maturing. This discipline, human factors or ergonomics, largely a product of World War II, looked at errors and accidents in quite a different light. Of course, people do things in unintended ways, but to label them human error and to stop at that contributes nothing to their elimination.
>
> In its infancy, this new discipline focused on the displays—the dials, gauges, indicators, printed materials from which people receive information about machine functioning—and on controls—knobs, levers, cranks, and push buttons people use to direct and issue commands to machines. Through redesign of these devices, human factors found that errors and accidents could, in many instances, be dramatically reduced.
>
> As it matured, human factors broadened its purview and now focuses on all the circumstances in which errors occur—the equipment, environment, procedures, users, skill levels, training, or generically, the system—and asks, "What is there about the system that allows a person to commit an error?" and "How could the system be changed or redesigned so that it would be difficult or impossible for even fallible humans to make mistakes?"

Attributing the harm done to a patient to the caregiver's (e.g., the anesthetist's) error, then, is a choice, a choice that locates the source of harm in the hands that were closest in space and time to the adverse event: the anesthetist's hands. We can stop once we have found those hands and direct all our interventionist zeal at them. We can remind the anesthetist to check oxygen saturation more often, to listen more carefully to alarms, and to intervene more aggressively when hearing them. Administrators can write a memo that tells all anesthetists to check manually for kinks in all tubing before every operation.

But that does not mean that there are no other choices, no other possible attributions. The anesthetist's hands, after all, are the place where all these other organizational, political, and ergonomic factors, pressures, constraints, and conflicts collect—factors that were all necessary and only jointly sufficient to produce the harm. The harm done to this patient can be attributed logically to any selection of them but only fairly to all of them together. When tracing back from the hands of the anesthetist, the causal web spreads quickly and widely, like cracks in a vandalized windowpane.

Human factors has evolved to take a radically different view of human error and the human contribution to accidents. It has distanced itself from the view that attributes trouble solely to unreliable human beings. The view that human factors has replaced is sometimes called "the old view" (American Medical Association

[AMA], 1998; Reason, 2000). This old view sees human error as a cause of failure and the primary target for intervention. This view has its roots in a period that is, as far as human factors is concerned, prehistoric.

Before World War II, practical and academic psychology was dominated by what was known as behaviorism. It deemed any study of why the mind did what it did as illegitimate and unscientific. What mattered was the tweaking with incentives and disincentives to get the mind to change its ways, never mind how that happened. Behaviorism focused on the overt signs of mental life only: the actions or inactions by the organism under study (whether that was a pigeon or a human). Through rewards and punishments, behavior was steered in the desired direction. Psychology assumed that the world in which people had to work was fixed. Humans had to adapt to its demands through selection and training (or, in medicine, through extraordinary and heroic competence, the topic of the next chapter). In this view of human error:

- Human error is the cause of adverse events.
- The systems in which people work are basically safe, or at least the system's weaknesses and risks are well known and good practitioners should guard against them. The chief threat to safety comes from the inherent unreliability of practitioners themselves.
- Safety can be improved by intervening at the level of the practitioner and his or her errors. These errors can be reduced through proceduralization, automation, training, discipline, and sanctions.

This view is not entirely foreign to healthcare even today. Failure in healthcare has been said to be the result of human ineptitude (Gawande, 2002), and thus healthcare needs individuals with the "strength of character to be virtuous" (Pellegrino, 2004). As Helmreich (2000) once opined in the *British Medical Journal*:

> Errors result from physiological and psychological limitations of humans. Causes of error include fatigue, workload, and fear, as well as cognitive overload, poor interpersonal communications, imperfect information processing, and flawed decision making. (p. 781)

Errors, in this view, are squarely a human problem, a problem of human limitations in thinking and talking. This old or prehistoric view easily becomes tautological. After all, do errors cause flawed decision making, or does flawed decision making cause errors? Or, *is* flawed decision making itself the error? To proponents, none of that matters. If the position is that only the outward signs of trouble need intervention, that errors need to be counted and repressed, then such tautologies are not fatal conceptual traps but simple, irrelevant technicalities. It is human unreliability (for whatever reason: fatigue, fear, overload, poor communication) that is the problem. This makes enhancing human reliability, by ensuring greater predictability and replicability of their behavior, the solution. This was once indeed the answer of the behaviorists.

With this position as a basis, however, accounts of error in healthcare can easily retreat into an old view. Take the story of three Colorado nurses who, by accidentally administering bezathine penicillin intravenously, caused the death of a neonate. The nurses were charged with criminal negligence, with one pleading guilty to a reduced charge. Another fought the charge and was eventually exonerated (Cook, Render, & Woods, 2000). Such an approach to adverse events actively seeks out the "bad apples" and assumes that with them gone, the system will be safer than before. The event is written off entirely to the error. And the error was made by those other deficient people. The emphasis on proximal causes ensures that the adverse event remains the result of a few uncharacteristically ill-performing individuals. They are not representative of the system or the larger practitioner population in it. Such a position leaves existing beliefs about the basic safety of the system intact. It often has economic or political motives:

> Formal accident investigations usually start with an assumption that the operator must have failed, and if this attribution can be made, that is the end of serious inquiry. Finding that faulty designs were responsible would entail enormous shutdown and retrofitting costs; finding that management was responsible would threaten those in charge, but finding that operators were responsible preserves the system, with some soporific injunctions about better training. (Perrow, 1984, p. 146)

But other attributions also are possible, even if they might become more expensive in the short term. In one study, 75% of adverse medication events were attributed to systemic factors rather than frontline errors (Gawande, Thomas, Zinner, & Brennan, 1999). Most of those involved in accident research and analyses are proponents of such attributions or of making multiple attributions. For example:

> Simply writing off ... accidents merely to [human] error is an overly simplistic, if not naive, approach. ... After all, it is well established that accidents cannot be attributed to a single cause, or in most instances, even a single individual. (Shappell & Wiegmann, 2001, p. 59)

Human factors has increasingly made attributions beyond single individuals likely and legitimate, not necessarily because they are true or truer than any other attribution. Given that attributions are judgments by people about something, they can neither be inherently accurate nor true. But they can be fair or unfair, and they can be productive or counterproductive. Pointing to the system rather than the individual has become recognized as both fairer and more productive. This was learned more than 70 years ago.

World War II brought such a furious pace of technological development that behaviorism was caught shorthanded. Practical problems emerged that were altogether immune to the behaviorist repertoire of motivational exhortations and incentives. Sustained operator vigilance in front of a radar screen was one such problem. Neither rewards nor punishments helped: Intervention at the level of the individual's actions or inactions proved useless. Workload management in visually and ergonomically extremely noisy environments was another such problem. Aviation, with its rapidly expanding use in all kinds of battle theaters and functions, and with some breathtaking revolutions in microprocessing, propulsion (e.g., the jet engine), range, communication, and navigation exerted great performance demands and generated a constant threat of fatigue, fear, and stress extremes. Aviation was at the forefront of the development of human factors and gave life to the assumptions that animate that field to this day. As recounted by Stanley Roscoe (1997), one of the eminent early engineering psychologists:

> It happened this way. In 1943, Lt. Alphonse Chapanis was called on to figure out why pilots and copilots of P-47s, B-17s, and B-25s frequently retracted the wheels instead of the flaps after landing. Chapanis, who was the only psychologist at Wright Field until the end of the war, was not involved in the ongoing studies of human factors in equipment design. Still, he immediately noticed that the side-by-side wheel and flap controls—in most cases identical toggle switches or nearly identical levers—could easily be confused. He also noted that the corresponding controls on the C-47 were not adjacent and their methods of actuation were quite different; hence C-47 copilots never pulled up the wheels after landing. (pp. 2–3)

The basis for the argument was laid. "Human error" was not some newly discovered psychological category of deficient human behavior. It was just a label, an attribution—a beginning, at best. It was a mere placeholder that said, "I don't really know what went wrong here," a placeholder that encouraged deeper probing, more investigation. Chapanis went behind the label to discover human actions that made perfect sense given the setting. He was even able to cross compare and show that a different engineered setting (the C-47) never triggered such actions. Human factors showed that the world was not fixed: Changes in the environment could easily lead to performance increments not achievable through behaviorist interventions. In behaviorism, performance had to be shaped after features of the world. In human factors, features of the world were shaped after the limits and capabilities of performance.

The complexity of engineered systems and organizations to manage them soared during World War II. It gave rise to the idea that the enemy of safety is not the unreliable human being but the complexity of the environment in which he or she is expected to work. The nuclear power industry in the West discovered this on the back of the Three Mile Island accident, during which the radioactive core of one of its two units melted down when it lost coolant in 1979 (Perrow, 1984). One particular human performance problem was the inability to recognize a loss-of-coolant accident when a relief valve stuck in the open position after doing its job, allowing coolant to escape. This problem could be traced back systematically to control room design and interface problems, which themselves were embedded in the enormous complexity of the plant and the processes of nuclear physics. As a result of the Three Mile Island incident, the industry enjoyed a human factors awakening in the 1980s.

Simplifying the operation or representation of system behavior can do wonders to improve system reliability and safety. The opposite is also true. A study in emergency care showed how safety improvements that added to people's task complexity hardly made things better or safer (Xiao et al., 1996). And the growth of complexity as the enemy of safety has been accelerating in healthcare over the last 30 to 40 years (Gawande, 2010; Woods, Patterson, & Cook, 2005).

Complexity in Healthcare: A Simple Human Factors Example

The number of prescription drugs available on the market has increased dramatically since 1980. Systems for packaging and distinguishing the vastly increased numbers of drugs, however, have hardly evolved in sync. Haavi Morreim (2004) picked it up from there:

> Consider a recently reported case in which an anesthesiologist, during surgery, reached into a drawer containing two vials, sitting side by side. Both had yellow labels and yellow caps. One was a paralytic agent and the other was a reversal agent to be used later, when the paralysis (for surgery) was no longer needed. At the beginning of the procedure, the doctor correctly administered the paralytic agent, but then toward the end of surgery he grabbed the wrong vial and instead of using the reversal agent, administered additional paralytic. No harm was done in this instance, but in discussing the episode with colleagues, the anesthesiologist found that many of them had committed precisely the same error. All knew of the hazard, yet none had spoken up about it. And so the situation inviting the error—identical vials sitting side-by-side in the same drawer—continued. (p. 215)

This, of course, is not just an ergonomic issue. There is a deeper organizational story about the possible reasons why practitioners may not have wanted to report making this error or why they did not flag the problem independent of whether they had been personally involved. Would it be fear of the consequences if they admitted something like this or a feeling of disinvolvement and lack of empowerment to change anything in the hospital or get anybody to listen? These topics are also within the purview of human factors (see Chapter 7) as constructive change to remove error potential from the system involves more than just the ergonomic insight that the error is possible.

Where the old view would see erratic, unreliable people with shortcomings in their thinking and their talking, human factors began to see and document patterns in human interaction with engineered devices and social settings, in situated cognition and people's collaborative work. Errors appeared to be systematically connected to features of people's tools, tasks, and operating environment. The human factors view, then, started seeing error not as a cause of failure but as a symptom of failure (Cook, 1998; Rasmussen & Batstone, 1989; Reason, 1997). Such a realization also can produce entirely different attributions for the causes of and remedies for iatrogenic harm. In the human factors view:

- Human error is a symptom of trouble deeper inside the system.
- The system is not basically safe, and the point is not to protect it from unreliable people. As noted by the World Health Organization (2005): "Every point in the process of care giving contains a certain inherent lack of safety: side-effects of drugs or drug combinations, hazards posed by a medical device, substandard or faulty products entering the health service, human shortcomings, or system failures." The system itself is full of contradictions between multiple goals that people must pursue simultaneously. People have to create safety.

- Human error is systematically connected to features of people's tools, tasks, and operating environment. Progress on safety comes from understanding and influencing these connections.

The human factors view of error represents a substantial movement for much of healthcare (Cook, 1998). It encourages the investigation of factors that easily disappear behind the label "human error" (e.g., system complexity, drug interactions, longstanding design problems). The rationale for this view is that human error is not an explanation for failure but instead demands an explanation. Effective countermeasures start not with individual human beings who themselves were at the receiving end of trouble much further upstream (Reason, 1997). As John Senders, one of the founding investigators of human error, once said, "Human error in medicine, and the adverse events which may follow, are problems of psychology and engineering not of medicine" (Woods et al., 2005, p. 2). The target for intervention, then, is the error-producing conditions present in their working environment. Indeed, for the World Health Organization (2005), the approach to patient safety *is* a human factors approach:

> Current conceptual thinking on the safety of patients places the prime responsibility for adverse events on deficiencies in system design, organization and operation rather than on individual providers or individual products. Similarly, most adverse events are not the result of negligence or lack of training, but rather occur because of latent causes within systems. For those who work on systems, adverse events are shaped and provoked by "upstream" systemic factors, which include the particular organization's strategy, its culture, its approach towards quality management and risk prevention, and its capacity for learning from failures. Countermeasures based on changes in the system are therefore more productive than those that target individual practices or products.
>
> Safety is a fundamental principle of patient care and a critical component of quality management. Its improvement demands a complex systemwide effort, involving a broad range of actions in performance improvement, environmental safety and risk management, including infection control, safe use of medicines, equipment safety, safe clinical practice and safe environment of care. It embraces nearly all health-care disciplines and actors, and thus requires a comprehensive, multifaceted approach to identifying and managing actual and potential risks to patient safety in individual services and finding broad long-term solutions for the system as a whole. Thinking in terms of "systems" offers the greatest promise of definitive risk-reduction solutions, which place the appropriate emphasis on every component of patient safety, as opposed to solutions driven by narrower and more specific aspects of the problem, which tend to underestimate the importance of other perspectives.

Let us pick up on the example of the nurse manager from the opening of the first chapter. The hospital where the nurse manager worked had recently embraced a state commitment by a local politician that surgical waiting lists should be shortened. (This had to be accomplished within 3 years because, a cynic would point out, that was the horizon in which the politician was re-eligible for office.) Nurses had been wooed away from the wards and into a newly built polyclinic that was going to conduct assembly-line minor elective operations in spiffy theaters with big windows and with no overnight stays. The concept succeeded: Surgical waiting lists slowly started to shrink. But where did the real costs collect? This happened on the wards. Replacement nurses were hard to find. Agency nurses were brought in to the extent possible, often from other countries. These nurses typically had no familiarity with other medical staff or with local routines, no inherent affinity to the hospital or its values, and little overlap with the cultural background (or even first language) of the patients. What happened to the likelihood of making mistakes? And would a nurse manager intent on weeding out the bad apples really dare to get rid of even more nurses?

One solution proposed in the hospital was to institute a "float nurse." This was a nurse employed in excess of staffing requirements, without a fixed workplace, who would drift to whatever ward was in acute need of more manpower. Predictably, regular ward staffing requirements were soon renegotiated downward. Because, after all, there was now an extra resource, right? This was a flexible one at that. The float nurse became the single source of slack that had to make up for cuts across the board. In short order, the position went from a nice-to-have backup to an absolutely essential resource who, in effect, had to be everywhere at all times—and was basically always too little, too late. These systemic factors conspired against, and were out of reach for, the individual competence of any nurse who had float duty.

ERROR AND EXPERTISE ARE TWO SIDES OF THE SAME COIN

Practitioners' strategies to cope with the peculiarities of the healthcare system in which they work are often successful. People in healthcare create safety through their practice. But they do not always succeed. Healthcare is laced with complexities, uncertainties, pressures, shortcomings, and contradictions between multiple goals that people must reconcile, deconflict, or pursue simultaneously. The openness of healthcare to political and economic interference makes it that policies, priorities, and funding can shift. There are continual developments in the organization and delivery of care, in staffing and personnel, in the use and complexity of medical technology and pharmaceuticals, as well as in procedures and routines. All of this can combine with production pressures and resource constraints to create new vulnerabilities and new forms of failure—even if it also creates new forms of economic and therapeutic success (AMA, 1998). New technology and new drugs, as well as new procedures and new management or even an entirely new polyclinic, all allow practitioners to be more successful at what they do. But these same things also create new pressure points, new gaps, new brittle areas, new skill and memory demands, and new failure modes.

This is the dual face of error and expertise. The things that make people good at delivering care also make them vulnerable to failure. This means that we cannot meaningfully separate the study of medical error from the study of normal human behavior inside the healthcare system (Rasmussen, 1985). As Ernst Mach (1976) said in 1905: "Knowledge and error flow from the same mental sources, only success can tell one from the other" (p. 84). The same organizational, operational, collaborative, and cognitive factors that allow people to be experts also govern the expression of error. Errors are not some mysterious product of the fallibility or unpredictability of people. Rather, errors are regular and predictable consequences of a variety of factors. In some cases, we understand a great deal about the factors involved, even in healthcare, while in others we currently know less (Woods, Dekker, Cook, Johannesen, & Sarter, 2010).

If error and expertise are both systematically connected to features of people, tools, tasks, and operating environment, then progress on safety comes from understanding and influencing those connections. The rationale is that human error is not an explanation for failure but instead demands an explanation. Effective countermeasures do not start with individual human beings who themselves were at the receiving end of much prior trouble.

> Simply writing off accidents merely to [human] error is an overly simplistic, if not naive, approach. After all, it is well-established that accidents cannot be attributed to a single cause, or in most instances, even a single individual. (Shappell & Wiegmann, 2001, p. 60)

The label *human error* really oversimplifies the role of human practitioners in creating both safe and unsafe outcomes. Even if the errors might seem ever so real to an observer, or to the practitioner, there are always other stories of what happened, of what went wrong—of what *is* wrong. Nurses may admit that they make mistakes, but those "mistakes" are only the beginning, only the first story. Behind the first story always lie multiple deeper stories. It is in those deeper stories that we may find the potential for constructive change.

Of course, human error can be seen as a problem of individual competence and as lying at the root of risk to patients. In that case, the human error problem is no more complex than getting rid of the incompetent individual or making that individual more competent. But it can also be seen as an organizational problem. Then the human error problem is at least as complex as the organization that helps produce it.

HUMAN ERROR AS ATTRIBUTION AND STARTING POINT

The nurse manager saw human error as a cause of trouble. Those "mistakes" she noted must have been a specific variety of human performance to her, something so clearly substandard and flawed that she would not be able to imagine how it could have been viewed as anything but substandard by the nurse at the time the act was committed. The wider, deeper story of staffing shortages, patient access and political coercion, resource constraints, and production pressures did not seem to enter the nurse manager's worldview. She elected to attribute the problem of patient safety to the mistakes of unreliable nurses. Perhaps it was the only attribution that made sense to her, that she felt should be within her remit, her control. Perhaps it was the only attribution she could afford. She might have needed it to sustain herself inside the hospital organization.

But saying that a nurse's mistakes are the cause of safety trouble is no more than a choice, a particular attribution. Indeed, human error is always an attribution. Research on what is now called "the fundamental attribution error" suggested that we are much more likely to attribute failures by other people to enduring and personal features of those whom we see as their cause (those nurses who make mistakes). When we fail in the same way, however, we tend to point to circumstances that prevented us from succeeding—factors outside ourselves. In that sense, the nurse manager's attribution was entirely consistent with the research. Of course, this makes her attribution no more or less "true" than any other attribution. The nurse manager chose to attribute her trouble to a nurse's mistakes and not to any other factors. She could also have chosen political interference as the source of trouble, or arrogant doctors, drug makers' refusal to design clearer labels and drop confusing names, or the sheer complexity of the system that delivers healthcare in her hospital.

That is not to say that the nurse manager's attribution does not make sense given her knowledge, her goals, and her mindset. In the wake of a failure, the pressure to do something can be so great, particularly on middle management, that they might be seduced to see human performance as puzzling, as perplexing, as the source of trouble. With the rubble of an adverse event on the ward spread before her, the manager can be forgiven for wondering why these nurses could not see what was obvious to her now. If only they had paid a little bit more attention. Something must be wrong with them. Perhaps those nurses need remediation. Perhaps they need disciplinary action to get them to try harder in the future. The nurse manager might feel the need to protect herself, her ward, her patients, and her organization against erratic and unreliable other people. And the nurse manager might feel that she needs to hold her nurses accountable for their mistakes and not allow them to blame the system, the organization.

But sticking with human error as attribution has been shown to be prejudicial and unspecific (Woods et al., 2010). Locating the source of trouble in single components at the sharp end has consistently retarded rather than advanced the understanding in organizations of how their processes fail. Labeling actions and assessments as "mistakes" identifies a symptom, not a cause (Hollnagel, Pedersen, & Rasmussen, 1981; Rasmussen, Duncan, & Leplat, 1987). It is not constructive to see people's mistakes as the cause of trouble. Rather, they are a symptom of deeper trouble, much of which has been around in the system for a long time (Table 2.1).

When people blame human error for their troubles (the nurses who make mistakes), they often make the implicit assumption that the system in which they work is

TABLE 2.1
Two Views of Human Error

Human Error as Medical Competence Problem	Human Error as Organizational Problem
Human error is a *cause* of trouble.	Human error is a *symptom* of trouble deeper inside the organization.
Human error can be the conclusion of an investigation.	Human error is a starting point for deeper investigation.
Human error is itself a useful target for intervention.	Meaningful intervention lies in the factors that help produce human expertise and error.
Healthcare is basically safe: It needs protection from unreliable humans.	Healthcare is not inherently safe. Only people can create safety by reconciling the multiple goals, pressures, constraints, and complexities.

already basically safe. Procedures, routines, and technologies are in place to deliver safe patient care, but until the people who work with it get their act together, such safety remains elusive. Decades of research into organizational failures show that when an adverse event happens, all the ingredients for it were already present in the system (Perrow, 1984; Reason, 1997; Turner, 1978). No new factors needed to be added to push the system over the edge into failure. Often, nothing uniquely stupid or egregious needed to be done. All that was needed was an unprecedented combination or concatenation of factors for their joint effect to emerge in a way that had not been seen or been produced before. This means that safety is not inherent in a healthcare system. The factors that produce risk to patients are always present, and the people who deliver care are often only the final hands through which accumulated successes and failures flow.

"I KNEW THIS COULD HAPPEN!"

Errors can seem so real. On discovering that we have made a mistake, even our own behavior may seem incredible. We wonder how on earth we could have missed this or that indication, how we possibly could have done this or omitted that. The nurse in the opening of Chapter 1, on learning that she had unintentionally hooked up an intravenous bag with the wrong drug in it, almost collapsed. She described how her legs would not carry her anymore, and that she felt she was about to faint. She felt so shocked and so disgusted with herself that she could hardly bear to look at her own hands, the hands that had hooked up the wrong bag and brought the patient harm. If this is how we can feel about our own actions, it is not surprising that a nurse manager (or a hospital director, a medical review board, or a prosecutor) may feel the same way.

THE HINDSIGHT BIAS

In hindsight, it may all be obvious. The data that we should have checked or noticed now jump out at us. If only we had seen that, then the mistake would never have happened. We would have done something differently. In hindsight, we clearly seem to know that this could happen. And we can hardly believe that we let it happen. This is known as the *hindsight bias* (Fischhoff, 1975). In hindsight, we know the outcome of a series of actions and events. We know how things turned out, and once we have this knowledge, we unwittingly start to straighten out and simplify the causal history that led up to the outcome. We amplify the significance of the data or the single action that could have steered things in a different direction. We tend to ignore the messiness of the context surrounding it, the many other things that were demanding our attention, and the fact that we actually did not know how things were going to turn out.

The hindsight bias has been well documented in psychological research and relates to how we perceive the probability of an event once we know that it has happened. As you might guess, once we know something has happened, we judge it likely to have happened and wonder why we did not foresee it. Baruch Fischhoff, who published the first article on this in 1975, did not actually use the term *hindsight bias.* Rather, he said, there is an unperceived, creeping determinism in our reasoning about cause once we know the outcome. Once we know the result, we tend to believe that the outcome was sure to happen, that everything was clearly pointing in its direction. It makes us believe that the outcome was foreseeable, and it makes us wonder how we did not foresee it. After all, in hindsight it is easy to point to the cues and indications that, if only we had noticed them, could have steered us to a different result.

The hindsight bias has significant consequences for what we consider real about human error. Recall that by extrapolating from local studies, the highly influential Institute of Medicine report concluded that between 44,000 and 98,000 Americans die each year as a result of medical errors (Kohn et al., 2000). Human error seemed to be a real and present danger. Even when using the lower estimate, deaths due to medical errors exceeded the number attributable to the eighth-leading cause of death. The report warned that more people died in a given year as a result of medical errors than from traffic accidents (43,458), breast cancer (42,297), or AIDS (16,516) (Kohn et al., 2000). In the following year, Hayward and Hofer (2001) questioned this suggestion that the number of deaths due to medical errors in hospitals is extremely high. They had 14 board certified, trained internists conduct 383 reviews of 111 hospital deaths, oversampling for markers previously associated with high rates of preventable deaths. Patients considered terminally ill who received palliative care only were excluded. Similar to previous studies, almost a quarter of patient deaths was rated as at least possibly preventable, with 6% rated as probably or definitely preventable.

After considering the 3-month prognosis, however, clinicians estimated that only half a percent of patients who died would have lived 3 months or more if care had been optimal. This represented roughly 1 patient per 10,000 hospital admissions. The study showed that the preventability of a patient's death is very much in the eye of the reviewer, and by extension, that patient death due to medical error is something that gets constructed largely in hindsight, after the patient has already died. With the benefit of hindsight, causality becomes simple. "People who know the outcome of a complex prior history of tangled, indeterminate events, remember that history as being much more determinant, leading 'inevitably' to the outcome they already knew" (Weick, 1995). Hindsight allows us to change past indeterminacy and complexity into order, structure, and oversimplified causality (Reason, 1990). Once the outcome of a dead patient is known, it is no longer so easily attributed to a huge confluence of events, factors, uncertainties, and contributions. With hindsight, the attribution human error becomes simple. The patient's death occurred because of misassessments, oversights, and other types of human mistakes.

Here is an example of how hindsight can drastically amplify the perceived likelihood of the outcome we now know about (Hugh & Dekker, 2009). A teaching hospital surgeon did a vagotomy (division of the vagus nerves to reduce acid secretion) and antrectomy (removal of the distal portion of the stomach) on a woman who had complained of pain and vomiting. Repeated endoscopy had revealed ulcers. Although the ulcer was healed, the symptoms persisted after the operation, and the woman was said to have developed paralysis of the stomach due to the vagotomy. She had not been warned about the possibility of this complication.

She later saw a gastroenterologist (who was not a surgeon), who suggested that the whole stomach should be removed. This was then done. She was crippled by nutritional problems after that and unable to work. The original surgeon was successfully sued for negligence. The court concluded that he had not done an on-table gastroscopy to see if the ulcer had healed. He had also failed to warn the woman of the estimated 1:10,000 chance of prolonged stomach paralysis. In hindsight, that chance seemed suddenly large and significant to the court—in large part because it had already happened. As for the on-table gastroscopy, it would not have changed the requirement for the operation. The surgeon appealed but was unsuccessful.

The hindsight bias also has strong implications for how we believe we should intervene in everyday practice. For example, if we find ourselves in the shoes of the manager whose nurses keep making those mistakes, hindsight means being able to look back, from the outside, on a sequence of events that led to an outcome that has already happened. It allows almost unlimited access to the true nature of the situation that surrounded people at the time (where they actually were versus where they thought they were; what state their system was in versus what they thought it was in). With hindsight, the entire sequence of events is exposed before us—the triggering conditions, its various twists and turns, the outcome, and the true nature of circumstances surrounding the route to trouble. Hindsight allows us to pinpoint what people missed and should not have missed; what nurses or employees did not do but should have done.

Hindsight biases our investigations or managerial intervention toward items that we now know were important. As a result, we assess people's decisions and actions mainly in the light of their failure to pick up these critical data. It artificially narrows our examination of the evidence and potentially misses alternative or wider explanations of people's behavior. Hindsight endows history, even immediate history, with a determinism it lacked while it was still unfolding, a determinism that comes from retrospective observers squeezing now-known events into the most plausible, convenient, or coherent deterministic scheme.

Our retrospective position contrasts fundamentally with the point of view of people who were inside the situation as it unfolded around them. To them, the outcome and the entirety of surrounding circumstances were not known. They contributed to the direction of the sequence of events on the basis of what they saw and understood to be the case on the inside of the evolving situation.

Look at Figure 2.1. You see an unfolding sequence of events. Such a linear sequence is not the only way to look at events, of course, as you will learn in Chapter 4. But here, for the sake of the explanation, the sequence has been given the shape of a tunnel meandering its way to an outcome. The figure shows two different perspectives on the pathway to failure. The first is the perspective from the outside and hindsight (typically our perspective after an adverse event has happened). From here, we can oversee the entire sequence of events—the triggering conditions, its various twists and turns, the outcome, and the true nature of circumstances surrounding the route to trouble. The other is the perspective from inside the tunnel. This is the point of view of people in the unfolding situation. To them, the outcome was not known (or they

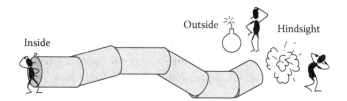

FIGURE 2.1   Different perspectives on a sequence of events: Looking from the outside and hindsight, you have knowledge of the outcome and dangers involved. From the inside, you may have neither.

would have done something else). They contributed to the direction of the sequence of events on the basis of what they saw on the inside of the unfolding situation.

Hindsight is baked deeply into the language of the stories on adverse events we tell one another. Take a common problem today—people losing track of the operation mode of their automated systems (e.g., in an infusion device). In hindsight, when we know how things developed and turned out, this problem is often called "losing mode awareness" or, more broadly, "loss of situation awareness." But this is perhaps little more than the difference between what we now know the situation actually was like and what people understood it to be at the time.

COUNTERFACTUAL REASONING

Tracing the sequence of events back from the outcome—which as outside observers we already know—we invariably come across junctures where people had opportunities to revise their assessment of the situation but failed to do so, where people were given the option to recover from their route to trouble but did not take it. These are counterfactuals—common in the analysis of adverse events. Counterfactuals prove what could have happened if certain minute and often utopian conditions had been met (see Figure 2.2).

Counterfactual reasoning may be a fruitful exercise when trying to uncover potential countermeasures against such failures in the future. But saying what people could have done to prevent a particular outcome does not explain why they did

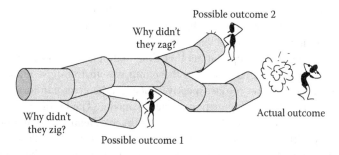

FIGURE 2.2   Counterfactuals: Going back through a sequence, you wonder why people missed opportunities to direct events away from the eventual outcome. This, however, does not explain failure.

what they did. This is the problem with counterfactuals. When they are enlisted as explanatory proxy, they circumvent the hard problem of understanding why people did what they did. Stressing what was not done (but if it had been done, the accident would not have happened) explains nothing about what actually happened or why. Counterfactuals are a powerful expression of the hindsight bias, as Starbuck and Milliken (1988) pointed out:

> Retrospective analyses always oversimplify the connections between behaviors and outcomes, and make the actual outcomes appear highly inevitable and highly predictable. Retrospection often creates an erroneous impression that errors should have been anticipated and prevented. (p. 337)

Counterfactuals impose structure and linearity on tangled prior histories. They can convert a mass of indeterminate actions and events, themselves overlapping and interacting, into a linear series of straightforward bifurcations. But human work in healthcare is seldom about simple dichotomous choices (such as to err or not to err). Bifurcations that yield clear previews of the respective outcomes at each end are rare. In reality, choice moments (such as there are) typically reveal multiple possible pathways that stretch out, like cracks in a window, into the ever-denser fog of futures not yet known. Their outcomes are indeterminate, hidden in what is still to come. In reality, actions need to be taken under uncertainty and under the pressure of limited time and other resources. What from the retrospective outside may look like a discrete, leisurely, two-choice opportunity not to fail is from the inside really just one fragment caught up in a stream of surrounding actions and assessments.

In fact, from the inside it may not look like a choice at all. These are often choices only in hindsight. To the people caught up in the sequence of events, there was perhaps not any compelling reason to reassess their situation or decide against anything (or else they probably would have) at the point the investigator has now found significant or controversial. They were likely doing what they were doing because they thought they were right, given their understanding of the situation and their pressures. The challenge for us becomes to understand how this may not have been a discrete event to the people whose actions are now controversial. We need to see how other people's decisions to proceed on their course of action were likely nothing more than continuous behavior—reinforced by their current understanding of the situation, confirmed by the cues they were focusing on, and reaffirmed by their expectations of how things would develop.

It is easy to show that people at another time and place did not know what you know today ("they should have known the pump was in free-flow mode"). When looked at from the position of retrospective outsider, the error can look so real, so compelling, but it is not an explanation of their behavior. The literature on the hindsight bias suggests that we should try to guard ourselves against mixing our reality with the reality of the people whose performance we are trying to understand. Those people did not know there was going to be a negative outcome, or they would have done something else. It is impossible for people to assess their decisions or incoming data in light of an outcome they do not yet know. As historian Barbara Tuchman (1981) put it: "Every scripture is entitled to be read in the light of the circumstances that brought it forth. To understand the choices open to people of another time, one must limit oneself to what they knew; see the past in its own clothes, as it were, not in ours" (p. 75). We need to try to see the evolving situation from the point of view of the people inside it and see why their assessments and actions made sense at the time (Dekker, 2006).

THE OUTCOME BIAS

Although often conflated with hindsight bias, the term *outcome bias* refers to the influence of outcome knowledge on the evaluation of decision quality. Remember that the hindsight bias also requires the outcome to be known but relates to retrospective estimates of the foreseeability of that outcome. The outcome bias means that we tend to think that if the outcome is bad, then the decision that led up to it also must have been bad.

The outcome bias has been demonstrated when clinicians make judgments about the appropriateness of care by other clinicians. And the worse the outcome is, the harsher the judgment will be. In fact, even the willingness of colleagues to make judgments about others' decisions increases when the outcome is bad. In the late 1980s, Robert Caplan and his colleagues studied a number of investigations of adverse anesthetic outcomes, collected from closed claims files of a nationwide group of U.S. professional liability insurance carriers. It revealed a statistically significant association between the severity of adverse outcomes and judgments of appropriateness of care. Nondisabling injuries were more often associated with a rating of appropriate care, while disabling injuries and death were more often associated with a rating of less-than-appropriate care. This suggested a strong presence of outcome bias: Highly unfavorable outcomes might predispose a reviewer toward harsher judgments, while minor injuries might elicit a less-critical response (Caplan, Posner, & Cheney, 1991).

To confirm whether a permanent injury would be more likely to elicit a rating of inappropriate care than a temporary injury, the authors then set up an experiment (Caplan, Posner, & Cheney, 1991). They asked more than 100 practicing anesthesiologists to judge the appropriateness of care in 21 cases involving adverse anesthetic outcomes. The original outcome in each case was classified as either temporary or permanent. They then generated a matching alternate case identical to the original in every respect *except* that a plausible outcome of opposite severity was substituted. The original and alternate cases were randomly divided across the reviewers, who did not know the aim of the study. Consistent with the preliminary analysis, they found a significant inverse relationship between severity of outcome and judgments of appropriateness of care in more than two of three cases. The proportion of ratings for appropriate care decreased by 31 percentage points when the outcome was changed from temporary to permanent and increased by 28 percentage points when the outcome was changed from permanent to temporary (Caplan et al., 1991). The experiment confirmed that knowledge of the severity of outcome influences a reviewer's judgment of the appropriateness of care.

WHY ARE THESE BIASES SO PERVASIVE?

The hindsight and outcome biases influence how we think about failure and about the role of human error. After the fact, we not only believe that the outcome was foreseeable and preventable but also are more willing to make negative judgments about people's decisions if the outcome was bad. The biases would have amplified the nurse manager's bewilderment about the mistakes made by her staff. If the bad outcomes were foreseeable and preventable, then why did her nurses not foresee and prevent them? There must be something wrong with the nurses who make mistakes, and she should get rid of them. Research has shown that both biases not only affect managers but also influence the views on medical error in expert reports, in assessments made by claims handlers, plaintiff and defense lawyers, disciplinary boards, tribunals, insurance company investigations, judges, and other judicial participants such as coroners and jurors. The hindsight bias is generally greater among plaintiff expert reports than in ones called by the defense. This is consistent with the specific commissioning of experts (either for or against) in adversarial systems and is known to be virtually unavoidable (Hugh & Dekker, 2009). In fact, asking people to be objective, to resist being influenced by their knowledge of outcome (e.g., by pretending that they do not know the outcome), is rarely successful. The hindsight and outcome biases seem indelibly encoded into our functioning as humans.

The outcome bias is consistent with our expectation that there is an equivalence between cause and consequence. If the effect is bad, then the cause must have been bad. This fits a basic Newtonian assumption about how the world works. Newton's third law of motion, after all, says that for every cause there is an equal and opposite effect. The effect is *equal* to the cause. We easily engineer this Newtonian physics concept into our social and organizational settings. Bad outcomes cannot happen without something that caused them. And with a bad outcome on our hands, we look for bad causes, such as the mistakes made by those nurses.

But in complex systems, bad outcomes can rarely be predicted on the basis of the functioning of separate constituent parts alone. All bad outcomes cannot be explained by the failure of such parts. Rather, bad outcomes can emerge even when everybody is doing good work, even when everybody follows the rules. This perspective is explained more fully in Chapter 7, which focuses on complexity theory and systems thinking and how it can inform us in new ways about the sources of success and failure in modern healthcare. Trivial, everyday organizational decisions, embedded in masses of similar decisions and only subject to special consideration with the wisdom of hindsight, cannot be meaningfully singled out because their relationship to the eventual outcome was complex and nonlinear and was almost impossible to foresee. Scott Snook (2000), who studied the mistaken shooting down of two U.S. helicopters by friendly forces over northern Iraq in 1993, expressed how this inability to find a bad cause for a bad effect "played with his emotions":

> This journey played with my emotions. When I first examined the data, I went in puzzled, angry, and disappointed—puzzled how two highly trained Air Force pilots could make such a deadly mistake; angry at how an entire crew of AWACS [Airborne Warning and Control System] controllers could sit by and watch a tragedy develop without taking action;

and disappointed at how dysfunctional Task Force OPC [Operation Provide Comfort} must have been to have not better integrated helicopters into its air operations. Each time I went in hot and suspicious. Each time I came out sympathetic and unnerved. ... If no one did anything wrong; if there were no unexplainable surprises at any level of analysis; if nothing was abnormal from a behavioral and organizational perspective; then what? (p. 203)

Snook's impulse to hunt down the cause (deadly pilot error, controllers sitting by, a dysfunctional task force) was doused by the lack of results. In the end, he came out "unnerved" because there was no "cause" that preceded the effect. This, of course, can leave managers and other accountable people empty handed, and they may still need to locate causes somewhere to feel able to do something about the problem. But in this, consequences cannot form the basis for an assessment of the gravity of the cause. There does not have to be proportionality in cause and consequence. In a complex system, bad consequences can happen without really bad assessments or decisions.

Acknowledging that bad outcomes arise from a hugely complex set of causes also can be difficult for the practitioners involved, particularly because of the hindsight bias. When we have been involved in an adverse event, we may quickly tell ourselves that we should have known, should have foreseen, should have noticed. If only we had zigged instead of zagged at a particular point in the unfolding events, then the outcome would have been different. Perhaps the hindsight bias is a highly adaptive, forward-looking, rational response to failure and may be more about predicting the future than explaining the past (Dekker, 2005; Hugh & Dekker, 2009). The linearization and simplification that we see happen with the hindsight bias may be a form of abstraction that allows us to export and project ours and others' experiences onto future situations.

Future situations can never be predicted at the same level of contextual detail as past situations. Predictions are possible only when we create a simple kind of model for the situations over which we wish to gain control. Exhaustively foreseeing every contextual factor, influence, or data point is impossible. The model we create when we oversimplify causality—naturally, effortlessly, automatically—after events with a bad outcome becomes one of binary choices, bifurcations, and unambiguous decision moments. That is perhaps the only useful model we can take into the future with us. It can guard us against the same type of pitfalls and offers quick guidance for coming forks in the road. But even if the hindsight bias is less about history than about the future and less about explaining than predicting and preventing, we should still try to be acutely aware of its influence on our understanding of adverse events.

JUDGING INSTEAD OF EXPLAINING

If, with the benefit of hindsight, an exit from the route to trouble stands out so clearly to outside observers, how was it possible for other people to miss it? If there was an opportunity to recover, to not harm a patient, to provide appropriate care, then failing to grab it or provide it demands an explanation. The place where observers often look for clarification is the set of rules, professional standards, and available data that surrounded people's operation at the time and how people did not see or meet that which they should have seen or met. Recognizing that there is a mismatch between what was done or seen and what should have been done or seen—as per those standards—we easily judge people for not doing what they should have done.

Where fragments of behavior are contrasted with written guidance that can be found to have been applicable in hindsight, actual performance is often found wanting; it does not live up to procedures or regulations.

A 2-year-old girl was killed in a U.S. hospital by a lethal dose of sodium chloride that had been mistakenly mixed into her chemotherapy bag. The pharmacist who approved the solution was convicted of involuntary manslaughter in connection with the girl's death. He was sentenced to spend 6 months in jail, followed by 6 months of house arrest and 3 years of probation. He was also ordered to pay a fine of $5,000 and spend 400 hours doing community service, during which he must seek pharmacological organizations to tell his story in the hopes that it would prevent others from making the same mistake.

The man was supervising pharmacist at Children's Hospital when one of his pharmacy technicians prepared a chemotherapy solution for the little girl. It was to be the last chemotherapy treatment before the patient was allowed to go home. The order called for a 1% saline base, but 3 days into the treatment, the solution instead came out with 23%. After awakening listless from an afternoon nap and complaining of a severe headache, the pain grew worse, and the girl began vomiting until she lost consciousness. She was rushed to the intensive care unit, where physicians tried to figure out what caused the complication. She slipped into a coma and remained on life support for several days, her brain so swollen that her eyes bulged. She was taken off life support and died a few days later.

"It was a senseless and preventable death," the mother read from a prepared statement in court. Turning to the pharmacist, she said: "You were the only person who could have prevented this from happening, and you didn't do it. You killed my baby." But was he? Did he? On the day the girl received the lethal solution, there had been a problem with the hospital's computer system, leaving the pharmacy with a backlog of drug orders and the workload to match it. The technician who actually mixed the solution had been reportedly preoccupied with planning her upcoming wedding. The explosion in pharmaceutical interventions and chemotherapy solutions available today was matched nowhere by a concomitant improvement in safe delivery techniques and practices. There was, as always, a whole system behind the apparently simple oversights and mistakes.

But according to the court, the explanation for the little girl's death was outrageous negligence on the part of the supervising pharmacist. The technician was not indicted or convicted; the hospital suffered no similar consequences. The only culprit was the pharmacist. The prosecutor pointed out that as supervising pharmacist, the defendant had the duty to inspect and approve all work prepared by technicians before the drugs were administered to patients, and this he did not do.

Investigations into adverse events can invest considerably in organizational archeology so that they can construct the regulatory or procedural framework within which the operations took place or should have taken place. Inconsistencies between existing procedures, guidance, or regulations and actual behavior are easy to expose when organizational records are excavated after the fact and rules uncovered that would have fit this or that particular situation.

This is not, however, informative. There is virtually always a mismatch located in hindsight between actual behavior and written guidance. Pointing out a mismatch sheds little light on the why of the behavior in question, and for that matter, mismatches between procedures and practice that are not unique to mishaps. There are also less-obvious or undocumented standards, like "good doctoring" (Gawande, 2002). These are often invoked when a controversial fragment knows no clear pre-ordained guidance but relies on local, situated judgment. For these cases, there are always supposed standards of good practice based on convention and putatively practiced across the entire industry. A lack of good doctoring (or something like "clinical judgment") can, if nothing else will, explain the variance in behavior that had not yet been taken into account.

When doing this, we frame peoples' past assessments and actions inside a world that we have invoked retrospectively. The problem is that this after-the-fact world may have little relevance to the actual world that produced the controversial behavior. That behavior is contrasted against the observer's reality, not the reality surrounding the behavior at the time. Judging people for what they did or did not do relative to some rule or standard does not explain why they did what they did. Saying that people failed to take this or that pathway—only in hindsight the right one—judges other people from a position of broader insight and outcome knowledge that they did not have. It does not explain a thing. It does not shed any light on why people did what they did given their surrounding circumstances. Outside observers have become caught in what William James called the "psychologist's fallacy" a century ago: They have substituted their own reality for the one of their object of study.

THE LOCAL RATIONALITY PRINCIPLE

What is striking about many adverse events is that people were doing exactly the sorts of things they would usually be doing—the things that usually led to success and safety. Adverse events are more typically the result of everyday influences on everyday decision making than they are isolated cases of erratic individuals behaving unrepresentatively. Adverse events are seldom preceded by egregious or inexplicable behavior. The challenge for patient safety is not why bad people produce adverse events but to understand why good people do. Rasmussen pointed out that if we cannot find a satisfactory answer to questions such as "how could they not have known?" then this is not because these people were behaving bizarrely. It is because we have chosen the wrong frame of reference for understanding their behavior (Vicente, 1999).

The frame of reference for understanding people's behavior should be their own normal work context, the context in which they were embedded and from whose point of view assessments and decisions were made. A challenge is to understand why assessments and actions that from the outside look like errors appear, from the inside, unremarkable, routine, normal, or systematically connected to features of the work environment in which people were placed. This is the local rationality principle: People are doing what makes sense given the situational indications, operational pressures, and organizational norms existing at the time.

HUMAN ERROR AS A RATIONAL CHOICE

The local rationality principle has grown out of a dissatisfaction with attempts to understand human functioning by reference to an ideal world in which people have access to all information all the time they need to reach a good decision. This was actually the dominant theoretical position well into the 1970s (Reason, 1990). *Rationalistic* means that mental processes can be understood with reference to normative theories that describe optimal strategies. Strategies may be optimal when the decision maker has perfect, exhaustive access to all relevant information, takes time enough to consider it all, and applies clearly defined goals and preferences to making the final choice. In such cases, errors are explained by reference to deviations from this rational norm, this ideal. If the decision turns out wrong, it may be because the decision maker did not take enough time to consider all information, or that he or she did not generate an exhaustive set of choice alternatives. Errors, in other words, are seen as deviant. They are departures from a standard. Errors are irrational in the sense that they require a motivational (as opposed to cognitive) component in their explanation. If people did not take enough time to consider all information, it is because they could not be bothered to do so. They did not try hard enough, and they should try harder next time, perhaps with the help of some training or procedural guidance. Investigative practice in healthcare is rife with such rationalist reflexes.

The assumption that errors are rational choices can be traced at least to the early 1800s, when human functioning was thought to follow the rules of the *Homo economicus*. The rational ideal was advanced again when psychologists embraced the computer as a model for human functioning in the 1950s and 1960s (Hollnagel, 2009). Computers were supposedly rational, after all. Their mechanistic, reasoned, and predictable form of decision making became the dominant metaphor for human functioning and remained so well into the 1970s.

The idealized decision maker meets a number of criteria (Hollnagel, 2009). The first is that the decision maker is completely informed: He or she knows all the possible alternatives and knows which courses of action will lead to which alternative. The decision maker is also capable of an objective, logical analysis of all available evidence on what would constitute the smartest alternative and is capable of seeing the finest differences between choice alternatives. Finally, the decision maker is fully rational and able to rank the alternatives according to their utility relative to the goals the decision maker finds important. These criteria were formalized in what was called subjective expected utility theory, which was devised by economists and mathematicians to guide human decision making (Reason, 1990). Its four basic assumptions were that people have a clearly defined utility function that allows them to index alternatives according to their desirability, that they have an exhaustive view of decision alternatives, that they can foresee the probability of each alternative scenario, and that they can choose among those to achieve the highest subjective utility.

Such idealized decision making of course requires a massive amount of cognitive resources, time, clearly defined values, and a good idea about the future. Only if a decision maker is infinitely wise, has infinite time, and has the luxury to chop decision problems into one-by-one considerations can he or she approximate something like the *Homo economicus*. This does not work in real life in healthcare or elsewhere.

If the starting point for explaining human behavior is a rationalist norm, then any decision that deviates from that norm can be explained only on the basis of irrationality or unawareness of the decision maker. In other words, if people do not behave formally according to utility theory or *Homo economicus*, it is because they either did not get all the relevant information together (they did not see the warning label on the intravenous bag, for example, and should have looked harder) or because they were irrational in their sorting and selecting the best decision alternative. Either way, people can be reminded to be more rational, to try harder, to be more motivated to do a reasonable job, and to follow the established norms that exist in their environment (e.g., clinical guidelines, procedures, checklists). The success of such interventions is not great. Telling all those other people to try harder is not going to make the human error problem go away (AMA, 1998; Woods et al., 2010).

The reason is the local rationality principle. People do not make decisions according to *Homo economicus* or utility theory. But people also do not come to work to do a bad job. Errors, as the other face of the expression of expertise, can better be understood as connected to the locally rational functioning of people inside a particular setting. There, multiple goals, knowledge and memory factors, and attentional dynamics all influence what people will consider rational right there and then (if they do so consciously or deliberately at all). This does not have to be globally or perfectly rational relative to some idealized decision maker with access to all relevant knowledge and decision alternatives. What matters is that it (mostly) works in that situation. Errors do not have to be explained by resorting to shortcomings in people's reason or motivation. Errors are the normal by-product of normal people doing normal work in resource-constrained, multigoal systems (Dekker, 2005).

ERROR AS THE BY-PRODUCT OF NORMAL WORK

Humans could not or should not even behave like perfectly rational decision makers. Whereas economists clung to the normative assumptions of decision making (decision makers have perfect and exhaustive access to information for their decisions, as well as clearly defined preferences and goals about what they want to achieve), psychology, with the help of artificial intelligence, posited that there is no such thing as perfect rationality (i.e., full knowledge of all relevant information, possible outcomes, relevant goals). There is not a single cognitive system in the world (neither human nor machine) that has sufficient computational capacity to deal with it all. From the 1970s onward, an increasing amount of psychological work pointed to the various ways in which human rationality is local, not global (and not "perfect" in that sense). Herbert Simon observed how the capacity of the human mind for formulating and solving complex problems is limited compared to the potential size of those problems, and that, as a result, human rationality is bounded. Objective (or perfect or global) rationality is cognitively impossible (in fact, it is also impossible for a computer too; decision problems always have more aspects and dimensions than can be enumerated in written code). As a result of this observation, Simon introduced the notion of "satisficing" (Newell & Simon, 1972). This is what people do: a combination of satisfying and sufficing. It captures how decisions in real-life situations are governed not by criteria of exhaustively computed utility but rather by whether the decision will work well enough for the problem at hand ("well" meaning "satisfy," and "enough" meaning "suffice").

Reasoning, these lines of research discovered, is governed by people's local understanding, by their focus of attention, goals, and knowledge rather than some global ideal. Human performance is embedded in, and systematically connected to, the situation in which it takes place. It can be understood (i.e., makes sense) with reference to that situational context, not by reference to some universal standard. Human actions and assessments can be described meaningfully only in reference to the localized setting in which they are made. They can be understood by intimately linking them to details of the context that produced and accompanied them. Such research has given rationality an interpretive flexibility. What is locally rational does not need to be globally rational. If a decision is locally rational, it makes sense from the point of view of the decision maker, which is what matters if we want to learn about the underlying reasons for what from the outside looks like error. The notion of local rationality removes the need to rely on irrational explanations of error. Errors make sense: They are rational, if only locally so, when seen from the inside of the situation in which they were made.

Real decision problems, then, resist the rationalistic format dictated for so long by economics and by extension, psychology. Options are not enumerated exhaustively. Access to information is incomplete at best, and people spend more time assessing and measuring situations than making decisions—if that is indeed what they do at all (Klein, 1993). In contrast to the prescriptions of the normative model, decision makers tend not to generate and evaluate several courses of action concurrently then to determine the best choice. People do not typically have clear or stable sets of preferences along which they can even rank the enumerated courses of action, picking the best one. Most complex decision problems actually do not have a single correct answer. Rather, decision makers in action tend to generate single options at the time, mentally simulate whether this option would work in practice and then either act on it or move on to a new line of thought (Klein, 1998). This is consistent with the notion of schemata: a mental representation of some aspect of the world that we can retrieve and with which we can effortlessly process situations and know how to act (Bartlett, 1932; Neisser, 1976). According to Bartlett, schemata are unconscious mental structures that reconstruct (rather than reproduce) earlier experiences. With a schema, it can be sufficient to recognize the situation that will trigger an appropriate response. This is the sort of decision making that works well when cognitive resources and time are limited. Wason and his colleagues did experiments in the early 1970s that showed how decision making is governed by *similarity matching* (which calls to mind Bartlett's schemata) more than by logic. Bruner showed that people prefer cues that have proved useful in the past, and thus had a "look of truth" about them. This, he said, was the criterion of verisimilitude, that made the wisdom of choosing some alternatives true enough, regardless of their present utility (Reason, 1990).

Schemata, and the sorts of decision making they enable, take the role of experience and expertise more seriously. By making use of what we have learned, there is no need to assess or compute our way through each situation from scratch (Hollnagel, 2009). What distinguishes good decision makers from bad decision makers, then, is foremost their ability to make sense of situations by using a highly organized experience base of relevant knowledge. Such reasoning about situations is more schema driven, heuristic, and recognition based than it is rational or computational. The typical decision setting does not allow the decision maker enough time or information to generate perfect solutions with perfectly rational calculations. Decision making in action calls for judgments under uncertainty, ambiguity, and time pressure. In those settings, options that appear to work are better than perfect options that never are computed. As long as the situation really is similar to what the schema suggests should be done, then this more effortless way of functioning in the world works well. It is a good heuristic, a good rule of thumb that gets applied automatically. But negative transfer can occur as well: that which has been found appropriate for one type of situation gets transferred to a situation that has some similarities but more important differences, which can lead to adverse outcomes (Hollnagel, 2009). Again, error and expertise are two sides of the same coin. What makes people efficient and accurate in some situations can lead to problems in others.

THE ACCOUNTABILITY BACKLASH

The local rationality principle says that what people do makes sense to them at the time; otherwise they would not be doing it. But we often have trouble with this. We keep discovering biases and aberrations in decision making. We see groupthink, confirmation bias, or routine violations of, for example, a hand-washing policy that seem hardly rational even from within a situational context. We insist that these deviant phenomena require motivational explanations and call for motivational solutions. People should be motivated to do the right thing; to pay attention; to double check; to sign discharge summaries, prescriptions, or operative notes; to read the label; to wash their hands. If they do not, then they should be reminded that it is their duty, their job, and perhaps they should face penalties if none of that appears to help. According to a proposal published in the *New England Journal of Medicine*, not practicing hand hygiene should trigger a penalty of education and loss of patient care privileges for 1 week, and not conducting a time-out before surgery should incur a penalty of 2 weeks' loss of patient care privileges (Wachter & Pronovost, 2009). It is as if there is an accountability backlash against the research results on local rationality.

Notice how easily this slips back into behaviorism. Through a system of rewards and punishments (threats of retribution), we hope to mold human performance after supposedly fixed features of the world. Wachter and Pronovost (2009, p. 276) suggested better policing of human behavior "with the use of methods such as video surveillance, computerized triggers, and unannounced, secret monitoring of compliance by hospital personnel." These initiatives suggest that the work environment is immune to further changes or improvements, and only that human non-compliance inside of it is the source of trouble. Better auditing and enforcement can make that trouble disappear. But what if these behaviors are just the effect, the outward signs, the symptoms, of deeper trouble as the work cited in this chapter strongly suggests?

Putatively motivational issues (such as deliberately breaking rules) must themselves be put back into context to see how human goals (getting the job done faster by not following all the rules to the letter) are made congruent with system goals through a collective of subtle pressures, subliminal messages about organizational preferences, and empirical success of operating outside existing rules. The point is not that people come to work to break the rules; they come to work to get a job done. The system in which they work wants fast turnaround times, maximization of capacity utilization, efficiency. Given those system goals (which are often kept implicit), rule breaking is not a motivational shortcoming but rather an indication of a well-motivated worker. Personal goals and system goals are harmonized, which in turn can lead to total system goal displacement: Efficiency is traded off against safety. If, for example, punitive pressure increases on hand washing while nothing is done to shorten the time and effort that patient rounds take, then token washing may become one way in which people will try to reconcile the various goals the system wants them to achieve simultaneously (Dekker & Hugh, 2010).

Discourse on patient safety often has trouble incorporating such subtle but powerful influences of organizational environments, structures, processes, and tasks into accounts of individual cognitive practices. In this regard, the field is conceptually underdeveloped. Indeed, how unstated cultural norms and values travel from the institutional, organizational level to express themselves in individual assessments and actions (and vice versa) is a concern central to sociology, which has only a small foothold in patient safety work. Bridging this macro-micro connection in the systematic production of errors and rule violations means understanding the dynamic interrelationships between issues as wide ranging as organizational characteristics and preferences, its environment and history, incrementalism in trading safety off against production, patterns and representations of safety-related information that are used as imperfect input to people's decision making, and the influence of hierarchies and production pressure and cultural dispositions on people's choices.

There is a lack of a language to cover these sorts of things in our discourse on patient safety. This book intends to fill part of that gap. The other hurdle to be overcome is the unyielding medical competency model. This also started Chapter 1: "If only I could get rid of the nurses who make mistakes." Of course, there is no substitute for medical experience, expertise, and competence and the deference and ethical responsibilities that come with them. Yet if we sustain the premise that medical competence or dedication is the only arbiter between failure and success in healthcare, then little progress can be made beyond forcing people to be more competent or motivated. The research base on safety, however, shows that much progress can be made beyond that limit. The relationships between people's situated competence and motivation on the one hand and their organizations, technologies, and human performance characteristics on the other can yield a rich trove of ways to improve patient safety. They are the topic of this book.

KEY POINTS

- Medical (in)competence is often seen as at the root of patient safety problems. But competence problems come from somewhere. They are the result of, for example, professional selection, medical education, skills training and maintenance, proficiency checking, and long-held cultural assumptions about infallibility and perfection in medicine.

- Competence alone rarely determines success or failure in healthcare. Many organizational, operational, and technological features that make medical work safe or unsafe lie beyond the reach of individual competence, such as waiting lists, bed shortages, organizational changes, production pressures, and technological error traps.
- "Human error" can be seen as the cause of trouble or as the symptom, the effect, of trouble deeper inside healthcare organizations. Much research since 1940 confirmed the latter. Human error is just an attribution, a label. It should be the starting point for a deeper investigation into the factors that lie behind the label.
- The hindsight bias relates to how we perceive the probability of an event once we know it has happened. Once we know the outcome of a series of events, we judge that outcome to be much more likely to happen than before we had that knowledge. Then we wonder why we did not foresee that outcome.
- The outcome bias makes us assume symmetry between cause and effect. If the outcome of a decision is bad, we quickly assume that the decision also was bad. Knowledge of the severity of outcome influences a reviewer's judgment of the appropriateness of care. Bad outcomes even make reviewers more willing to make judgments of others' performance.
- The local rationality principle says that what people do makes sense to them given the goals they are pursuing, the knowledge they have available about the problem, and the direction of their attention at the moment. People do not come to work to do a bad job. It has a long history in research on human performance and decision making. Despite this, there are still often calls for medical practitioners to be more competent or motivated—to try harder.

REFERENCES

American Medical Association. (1998). *A tale of two stories: Contrasting views of patient safety.* Chicago: Author.

Baker, G. R., Norton, P. G., Flintoft, V., Blais, R., Brown, A., Cox, J., et al. (2004). The Canadian Adverse Events Study: The incidence of adverse events among hospital patients in Canada. *Canadian Medical Association Journal, 170*(11), 965–968.

Bartlett, F. (1932). *Remembering: An experimental and social study.* Cambridge, UK: Cambridge University Press.

Berwick, D. M., Calkins, D. R., McCannon, C. J., & Hackbarth, A. D. (2006). The 100,000 lives campaign: Setting a goal and a deadline for improving health care quality. *Journal of the American Medical Association, 295*(3), 324–327.

Berwick, D. M., Hackbarth, A. D., & McCannon, C. J. (2006). IHI replies to "The 100,000 Lives Campaign: A scientific and policy review." *Joint Commission Journal on Quality and Patient Safety/Joint Commission Resources, 32*(11), 628–630; discussion 631–623.

Caplan, R. A., Posner, K. L., & Cheney, F. W. (1991). Effect of outcome on physician judgments of appropriateness of care. *Journal of the American Medical Association, 265*(1957–1960).

Chapanis, A. (2004). Foreword. In M. S. Bogner (Ed.), *Misadventures in health care: Inside stories* (pp. xi–xiv). Mahwah, NJ: Erlbaum.

Cook, R. I. (1998). *Two years before the mast: Learning how to learn about patient safety.* Paper presented at Enhancing Patient Safety and Reducing Errors in Health Care: A Multidisciplinary Leadership Conference, November 1998, Rancho Mirage, CA.

Cook, R. I., Render, M., & Woods, D. D. (2000). Gaps in the continuity of care and progress on patient safety. *British Medical Journal, 320*(7237), 791–795.

Davis, P., Lay-Yee, R., Briant, R., Ali, W., Scott, A., & Schug, S. (2002). Adverse events in New Zealand public hospitals I: Occurrence and impact. *The New Zealand Medical Journal, 115*(1167), U271.

Dekker, S. W. A. (2005). *Ten questions about human error: A new view of human factors and system safety.* Mahwah, NJ: Erlbaum.

Dekker, S. W. A. (2006). *The field guide to understanding human error.* Aldershot, UK: Ashgate.

Dekker, S. W. A., & Hugh, T. B. (2010). Balancing "no blame" with accountability in patient safety. *New England Journal of Medicine, 362*(3), 275.

Department of Health. (2000). *An organization with a memory: The report of an expert group on learning from adverse events.* London: Her Majesty's Stationery Office.

Fischhoff, B. (1975). Hindsight ≠ foresight: The effect of outcome knowledge on judgment under uncertainty. *Journal of Experimental Psychology: Human Perception and Performance, 1*(3), 288–299.

Gawande, A. (2002). *Complications: A surgeon's notes on an imperfect science.* New York: Picado.

Gawande, A. (2010). *The checklist manifesto: How to get things right.* New York: Metropolitan Books.

Gawande, A., Thomas, E. J., Zinner, M. J., & Brennan, T. A. (1999). The incidence and nature of surgical adverse events in Colorado and Utah in 1992. *Surgery, 126*(1), 66–75.

Golden-Biddle, K., & Locke, K. (1993). Appealing work: An investigation of how ethnographic texts convince. *Organization Science, 4*(4), 595–616.

Hayward, R. A., & Hofer, T. P. (2001). Estimating hospital deaths due to medical errors: Preventability is in the eye of the reviewer. *Journal of the American Medical Association, 286*(4), 415–420.

Helmreich, R. L. (2000). On error management: Lessons from aviation. *British Medical Journal, 320*(7237), 781–785.

Hollnagel, E. (2009). *The ETTO principle: Efficiency-thoroughness trade-off. Why things that go right sometimes go wrong.* Aldershot, UK: Ashgate.

Hollnagel, E., Pedersen, O. M., & Rasmussen, J. (1981). *Notes on human performance analysis.* Roskilde, Denmark: Risø National Laboratory.

Hugh, T. B., & Dekker, S. W. A. (2009). Hindsight bias and outcome bias in the social construction of medical negligence: A review. *Journal of Law and Medicine, 16*(5), 846–857.

Klein, G. A. (1993). *Decision making in action: Models and methods.* Norwood, NJ: Ablex.

Klein, G. A. (1998). *Sources of power: How people make decisions.* Cambridge, MA: MIT Press.

Kohn, L. T., Corrigan, J., & Donaldson, M. S. (2000). *To err is human: Building a safer health system.* Washington, DC: National Academy Press.

Leape, L. L., Brennan, T. A., Laird, N., Lawthers, A. G., Localio, A. R., Barnes, B. A., et al. (1991). The nature of adverse events in hospitalized patients: Results of the Harvard Medical Practice Study II. *New England Journal of Medicine, 324*(6), 377–384.

Mach, E. (1976). *Knowledge and error: Sketches on the psychology of enquiry.* Dordrecht, Netherlands: Reidel.

Morreim, E. H. (2004). Medical errors: Pinning the blame versus blaming the system. In V. A. Sharpe (Ed.), *Accountability: Patient safety and policy reform* (pp. 213–232). Washington, DC: Georgetown University Press.

National Safety Council. (2004). *Injury facts 2004 edition.* Itasca, IL: National Safety Council.

Neale, G., Woloshynowych, M., & Vincent, C. (2001). Exploring the causes of adverse events in NHS hospital practice. *Journal of the Royal Society of Medicine, 94*(7), 322–330.

Neisser, U. (1976). *Cognition and reality: Principles and implications of cognitive psychology*. San Francisco: Freeman.

Newell, A., & Simon, H. A. (1972). *Human problem solving*. Englewood Cliffs, NJ: Prentice-Hall.

Paget, M. A. (2004). *The unity of mistakes: A phenomenological interpretation of medical work*. Philadelphia: Temple University Press.

Pellegrino, E. D. (2004). Prevention of medical error: Where professional and organizational ethics meet. In V. A. Sharpe (Ed.), *Accountability: Patient safety and policy reform* (pp. 83–98). Washington, DC: Georgetown University Press.

Perrow, C. (1984). *Normal accidents: Living with high-risk technologies*. New York: Basic Books.

Rasmussen, J. (1985). Trends in human reliability analysis. *Ergonomics, 28*(8), 1185–1195.

Rasmussen, J., & Batstone, R. (1989). *Why do complex organizational systems fail?* Washington, DC: World Bank.

Rasmussen, J., Duncan, K., & Leplat, J. (1987). *New technology and human error*. Chichester, UK; Wiley.

Reason, J. T. (1990). *Human error*. New York: Cambridge University Press.

Reason, J. T. (1997). *Managing the risks of organizational accidents*. Aldershot, UK: Ashgate.

Reason, J. T. (2000). *Grace under fire: Compensating for adverse events in cardiothoracic surgery*. Paper presented at the Fifth Conference on Naturalistic Decision Making May 2000, Stockholm, Sweden.

Roscoe, S. N. (1997). The adolescence of engineering psychology. In S. M. Casey (Ed.), *Volume 1, Human factors history monograph series*. Santa Monica, CA: Human Factors and Ergonomics Society.

Shappell, S. A., & Wiegmann, D. A. (2001). Applying reason: The human factors analysis and classification system. *Human Factors and Aerospace Safety, 1*, 59–86.

Snook, S. A. (2000). *Friendly fire: The accidental shootdown of U.S. Black Hawks over northern Iraq*. Princeton, NJ: Princeton University Press.

Starbuck, W. H., & Milliken, F. J. (1988). *Challenger*: Fine-tuning the odds until something breaks. *The Journal of Management Studies, 25*(4), 319–341.

Thomas, E. J., Studdert, D. M., Burstin, H. R., Orav, E. J., Zeena, T., Williams, E. J., et al. (2000). Incidence and types of adverse events and negligent care in Utah and Colorado. *Medical Care, 38*(3), 261–271.

Thomas, E. J., Studdert, D. M., Runciman, W. B., Webb, R. K., Sexton, E. J., Wilson, R. M., et al. (2000). A comparison of iatrogenic injury studies in Australia and the USA. I: Context, methods, casemix, population, patient and hospital characteristics. *International Journal for Quality in Health Care, 12*(5), 371–378.

Tuchman, B. W. (1981). *Practicing history: Selected essays*. New York: Knopf.

Turner, B. A. (1978). *Man-made disasters*. London: Wykeham.

Vicente, K. J. (1999). Cognitive work analysis: Toward safe, productive, and healthy computer-based work. Mahwah, NJ: Erlbaum.

Vincent, C., Neale, G., & Woloshynowych, M. (2001). Adverse events in British hospitals: Preliminary retrospective record review. *British Medical Journal, 322*(7285), 517–519.

Wachter, R. M., & Pronovost, P. J. (2006). The 100,000 Lives Campaign: A scientific and policy review. *Joint Commission Journal on Quality and Patient Safety/Joint Commission Resources, 32*(11), 621–627.

Wachter, R. M., & Pronovost, P. J. (2009). Balancing "no blame" with accountability in patient safety. *New England Journal of Medicine, 361*, 1401–1406.

Weick, K. E. (1995). *Sensemaking in organizations*. Thousand Oaks, CA: Sage.

World Health Organization. (2005). *World alliance for patient safety: Forward program 2005*. Geneva: World Health Organization.

Woods, D. D., Dekker, S. W. A., Cook, R. I., Johannesen, L. J., & Sarter, N. B. (2010). *Behind human error.* Aldershot, UK: Ashgate.

Woods, D. D., Patterson, E. S., & Cook, R. I. (2005). *Behind human error: Taming complexity to improve patient safety.* Columbus, OH: Institute for Ergonomics, The Ohio State University.

Xiao, Y., Hunter, W. A., Mackenzie, C. F., Jeffries, N. J., Horst, R. L., & Group, L. (1996). Task complexity in emergency medical care and its implications for team coordination. *Human Factors, 38,* 636–645.

3 Cognitive Factors of Healthcare Work

What is now known as "the human factors approach" was once characterized by a focus on individual people and their functioning. In the 1970s, human factors mostly focused on models of mental information processing. These models assumed that the interesting activities for understanding functioning in complex worlds were internal to the human. Meaning and perceptual order are the end result of an internal trade in representations. These representations would get increasingly filled out and meaningful as a result of processing in the mind, with the help of structures like short-term memory, long-term memory, decision making, action execution, and others. Information processing fit a larger, dominant meta-theoretical perspective that takes the individual as its central focus. This view was a heritage of the scientific revolution, which increasingly popularized the humanistic idea of a "self-contained individual." For most of human factors, this meant that all processes worth studying took place within the boundaries of the body (or mind), something epitomized by the mentalist focus of information processing.

This is quite different today. The unit of analysis for human factors is now the human in the context of other people, the organization, and technical artifacts associated with their work. No models of only individual cognition in the head, after all, can be authentic to how practitioners accomplish work in safety-critical settings. Activities such as decision making, situation assessment, and even memory are not the private mental domain of minds closed off to the world. They are distributed across other people and artifacts. Situation assessment and decision making in many situations are a team activity, supported in technology of various kinds. Memory is something we put in the world in the form of devices, notes, and other artifacts. Of course, all of this is embedded in a larger organization that helps set goals, constraints, and opportunities for those carrying out safety-critical work.

The framework used for this chapter is based on this (Woods, Dekker, Cook, Johannesen, & Sarter, 2010). It does not contain specific, distinct models of internal cognitive mechanisms. Instead, it offers a guide to the cognitive functions that people must perform, in concert with other people and technical artifacts, to handle the demands of complex fields of practice. It proposes that cognitive work in complex worlds involves attentional dynamics (where to focus attention, when, at the cost of what else), knowledge factors (how to learn, store, activate, and deploy knowledge), and strategic factors (how to deal with larger constraints and goals imposed by the organization in which the work is carried out). The discussion of these factors can be seen as selective rather than comprehensive, and the chapter points to other literature that deals with them in more detail.

ATTENTIONAL DYNAMICS

In cognitively noisy worlds, people routinely need to keep track of multiple threads of activity. How do they know where to focus when? How does the allocation of attention work? People update their understanding of an unfolding situation based on incoming cues. This understanding in turn directs them to act (or not) in one way or another, which changes the situation (according to expectations or not), which in turn updates people's understanding of what is going on. This is known as Neisser's circle (Neisser, 1976) (see Figure 3.1). It shows how sense making is ongoing. People's actions and assessments of what is going on are deeply intertwined. By doing something, people generate more information about the world. This in turn helps them decide which actions to take next. Action and situation assessment are tightly interwoven: One constrains and informs the other.

COGNITIVE FIXATION AND VAGABONDING

In dynamic situations, people direct their attention as a joint result of their current understanding of the situation, their knowledge and goals, and of what happens around them. Current understanding helps people form expectations about what should happen next (either as a result of their own actions or as a result of changes in the world itself). Particularly salient or intrusive cues will draw attention even if they fall outside people's current interpretation of what is happening. Expertise helps direct attention even before things happen. Yan Xiao and colleagues studied planning expertise in anesthesiology through observation of 40 cases (Xiao, Milgram, & Doyle, 1997). It appeared that expert anesthetists search for cues or warning signs that may indicate potential problems and provide information on obstacles to potential solutions. Much of this is done before the time of surgery so that they know where to look and what to look out for even before beginning surgery. That makes good sense because time and opportunity for discovery and repair may be significantly reduced once a patient has gone into surgery.

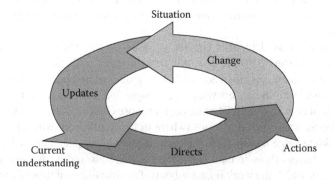

FIGURE 3.1   Neisser's cognitive cycle. We make assessments about the world, updating our current understanding. This directs our actions in the world, which change what the world looks like, which in turn updates our understanding, and so forth.

Keeping up with a dynamic world, in which situations evolve and change, is a demanding part of much medical work. People may fall behind in rapidly changing conditions, and update their interpretation of what is happening constantly, trying to follow every little change in the world. Or people become locked in one interpretation, even while evidence around them suggests that the situation has changed. This is known as *cognitive fixation*, or *cognitive lockup* (Cook, McDonald, & Smalhout, 1989). De Keyser and Woods (1990) have described several patterns of practitioner fixation. One of those was "this and nothing else," by which practitioners are stuck on one strategy, one goal, and seem unable to consider other possibilities. There can be a great deal of persistence in this kind of fixation. Practitioners may repeat an action or recheck the same data channels several times, for example. This pattern is easy to see because of the unusual level of repetitions despite an absence of results. Even if they understand the absence of results, they do not change in strategy. Another pattern is "everything is okay." In this case, the practitioners do not react to the change in the environment even if, in hindsight, evidence was available that something was going wrong or different. Such evidence may be discounted or rationalized away (and often with good reason; expertise and experience may tell the practitioner that these things are usually red herrings or false alarms).

Characteristic of cognitive fixation is that the immediate problem-solving context biases people in some direction (e.g., "this is an indication problem"). From an emerging mass of uncertain, incomplete, and contradictory data, people have to come up with a plausible explanation, an explanation that covers at least part of the data observed. But it can activate certain kinds of knowledge and troubleshooting activities at the expense of others. These early conjectures can become assumptions that are taken for granted in subsequent diagnosis and problem solving. This has been called *plan continuation* (Orasanu & Martin, 1998). Situational dynamics and the continuous emergence of incomplete, uncertain evidence play a role. Early cues that suggest the initial plan is correct are usually strong and unambiguous when people become locked into a continuation of their plan. Later cues that suggest the plan should be abandoned are typically fewer, more ambiguous, and weak. These cues, even while people see them and acknowledge them, often do not succeed in pulling people in a different direction. Also, conditions may deteriorate gradually, all but hiding the evidence of decline.

The challenge, of course, is to understand why it made sense to people to continue with their original plan. Which cues did they rely on, when, and why? What were the contradictions and ambiguities raised by the multiple cues and by the order in which they emerged? Abandoning the plan also may be costly on various dimensions, which must be weighed against the weakness and limited number of the cues that speak for it. Going from laparoscopic to open surgery entails multiple organizational, clinical, and economic consequences, for example. It involves new risks for the patient. People may need a lot of convincing evidence to justify changing their plan in such a case. This evidence may not be compelling until afterward.

Fixation is likely one extreme end of what is a cognitive continuum, and the skewed result of a cognitive balancing act. In cognitively noisy or diagnostically difficult situations, not every cue or indication is important. Some coherence and stability in explanation of what might be going on is necessary to be able to act at all.

If no coherence or stability is present, then the opposite may occur: thematic vaga-bonding. Here, the practitioner flits from one explanation to another with every new cue that comes in, never resting to form a coherent story of what might be happening. Their behavior looks incoherent because they are often jumping from one reaction to another without much success. Many hypotheses are generated, but never a whole or correct one. The fixation-vagabonding continuum, then, suggests that revisions of situation assessment can come too quickly or too slowly relative to how the underly-ing problem is unfolding.

The cognitive balancing act involves various activities that may go well or wrong, again confirming how error and expertise are two sides of the same coin. Shifting and scheduling attention as the incident unfolds can be too rapid or too slow, too disintegrated or too fixed. Knowledge needs to be brought to bear (see next section) but simply may not come to mind in that context because the way knowledge is orga-nized can be ill matched to the unfolding situation. Additional hypotheses may not be generated because a strong or familiar one (which might still be wrong) is retained as it usually works very well. Single-factor assessments of what is going on may have to be ditched in favor of the probable contribution of multiple factors to the incident evolution. Thus, any problems that occur should be attributed to the interaction of particular environmental and task features, and the heuristics people apply, rather than to any bias in the strategies used. The way that a problem presents itself to prac-titioners may make it easy to entertain plausible but in fact erroneous possibilities.

The activities that help produce fixation or vagabonding also lay out ways for intervention, for breaking out of counterproductive attentional processes in criti-cal situations. Research consistently shows that revising assessments successfully requires a new way of looking at previous facts. It takes a fresh perspective (Woods, 1988; Woods & Patterson, 2000). This can be done, for instance, by bringing people in who are new to the situation (of course, this increases coordination demands, as mentioned in the next section, but it may be worth it). It is important that these people were not part of the initial problem formulation and were thus not captured by any fixation that might have occurred. If new people cannot be brought in, other groups can be consulted that have diverse knowledge or tool sets. Even new types of display or representation design can help put data in perspective, which can reveal erroneous assumptions or hypotheses, or refocus the team on what is most important in that scenario.

ESCALATION AND DYNAMIC FAULT MANAGEMENT

Another aspect of managing dynamic problems is that people have to commit cog-nitive resources to solving them *while* maintaining process integrity (e.g., patient ventilation). This is called *dynamic fault management* and is typical for event-driven domains. Figuring out what is wrong has to be done while keeping the process going (e.g., keeping a patient alive). Diagnosis and maintaining process integrity are often closely intertwined; not doing troubleshooting or corrective activities may well chal-lenge the integrity of the entire process.

As situations escalate, the criticality, workload, and coordination demands all increase. This complicates problem solving and the maintenance of process integrity, as well as their interrelationship (see the explanation of "tight coupling" in Chapter 5). Escalating situations are characterized by a cascade of problem effects, an increase in tempo, and most often, a scenario that wanders outside canonical or textbook cases. Things happen in quicker succession, perhaps seemingly unrelatedly, margins for recovery may decrease, and the criticality of the situation increases (Woods & Patterson, 2000).

Escalation makes additional demands on cognitive activity. The need for knowledge mounts as effects become visible. There are more information sources to monitor. This in turn swells the data set that needs to be integrated into a coherent assessment of what is occurring. More needs to be reviewed; more may need to be revised as these new data are received. Actions to protect the integrity and safety of systems need to be identified, carried out, and monitored for success. Existing plans need to be modified or new plans formulated to cope with the consequences of lack of congruency between plans and reality. Contingencies need to be considered. Multiple threads of activity, tasks, and possible hypotheses easily disrupt practitioners' control of their attention. Such situations are naturally more vulnerable to fixation, vagabonding, or plan continuation. Resilience against the corrosive effects of escalation can be created when practitioners engineer some extra margin into their operations to give them the kind of slack that allows them to adapt under pressure and accommodate surprise and to extemporize and accommodate the fluctuating pressures and task loads of actual work.

Escalation also increases coordination demands across people. Knowledge may reside in different people, may be part and parcel of certain specialties, or can be found only in different parts of the system. Multiple people or groups may have to coordinate to implement activities aimed at gaining information to aid diagnosis or to protect the monitored process. The trouble in the underlying process requires informing and updating others who may be affected by the consequences or who might help with recovery.

Contingency theory explains how organizations adapt their structure and internal workings after the conditions (indeed, contingencies) they meet in their environment (Mintzberg, 1979). Escalating situations do predictable things with the organization and coordination of work. The more dynamic the environment, the more organic organizational structures become. Responding to rapidly changing local conditions requires decentralization. Information should not pass up and down decision ladders because it takes too much time and capacity; such channels clog up quickly. Chapter 6 highlights the importance of diversity and flatter communication in problem-solving teams, particularly under conditions of uncertainty, tight coupling, and interactive complexity (Weick & Sutcliffe, 2007).

MULTITASKING AND PROSPECTIVE MEMORY

Practitioners often do not have the luxury of deferring one task until another is completed. They will have to interleave the execution of multiple tasks at the same time. This is commonly known as *multitasking*. Multitasking involves the constant switching of attention between the different tasks and the accomplishment of steps of each in alternation with the others. When tasks are practiced together consistently and frequently, practitioners become able to perform them simultaneously as if they were one task—or in such a way that one task demands little, if any, conscious attention (Loukopoulos, Dismukes, & Barshi, 2009).

But clinicians will encounter many situations in which multiple tasks need to be accomplished, and multiple threads of attention kept track of, that are not as rehearsed. Such demands on human performance make it vulnerable to errors. Becoming absorbed in one task at the expense of others is one of the risks. Even though practitioners might tell themselves to keep checking on the other tasks, being absorbed in one task often erodes cognitive timekeeping (Hancock & Weaver, 2005). In other words, practitioners may not realize how long they have been away from attending to the other tasks or attentional threads and might be surprised to learn about the length after pulling out of the one task on which they were so focused. This is where well-communicating teams might be less vulnerable than individuals: Multiple people can help in the allocation of attention to those multiple threads (see Chapter 7).

Allocating attention and switching between tasks demands cognitive resources (either from the team or from the individual). It is a kind of mental bookkeeping—keeping track of what has already been accomplished and what has still been left undone across multiple different tasks or strands of activity. This often involves what is known as prospective memory: remembering to do something in the future. As Key Dismukes (2006) of NASA (National Aeronautics and Space Administration) explains:

> Prospective memory is distinguished by three features: (1) an intention to perform an action at some later time when circumstances permit, (2) a delay between forming and executing the intention, typically filled with activities not directly related to the deferred action, and (3) the absence of an explicit prompt indicating that it is time to retrieve the intention from memory—the individual must "remember to remember." ... Typically, if queried after forgetting to perform an action, individuals can recall what they intended to do. Thus the critical issue in prospective memory is not retention of the content of intentions but retrieval of those intentions at the appropriate moment, which is quite vulnerable to failure. (p. 2242)

Also known as remembering to remember, the success of prospective memory often depends on the availability of cues that call the practitioner's attention to the task or step to be accomplished. If the cue is based on some sort of event in the environment, then this might not be so hard. The practitioner may tell her- or himself to do or check something when something else happens. There might be no need to attend to the other task until or unless such an event occurs and represents the cue or trigger for prospective memory to activate. These external cues may in fact arise from another concurrent task in which the practitioner is engaged, with one task prompting the practitioner to attend to another. But in many cases, prospective memory needs to rely on time, with the conditions for performing the deferred task defined in terms of time passed (e.g., "I will check blood pressure every 2 minutes").

The cognitive demands of time-based prospective memory may differ significantly from those imposed on event-based prospective memory (Loukopoulos et al., 2009). In the former, deferred intention is not associated with specific external cues and may need to rely on internal timekeeping, something that suffers under the pressure of focused attention on a particular task (Hancock & Weaver, 2005). But any kind of prospective memory is vulnerable to a variety of events in practitioners' work that can disturb the relationship between intention and action. These include the events discussed next (Loukopoulos et al., 2009).

First are interruptions and distractions that occur and divert attention from an ongoing task (or set of tasks), causing it to be suspended at least temporarily. In most clinical practice, interruptions are so common that they might be regarded as unremarkable and not associated with heightened risk or vulnerability. There is an interesting distinction between interruptions and distractions: Interruptions can be thought of as discrete events (some alarm going off or somebody demanding attention) that need attention immediately. Some interruptions can be predictable and frequent, and practitioners might develop action scripts for dealing with them swiftly and without sacrificing much in the way of cognitive resources. Distractions can be lengthier and pull attentional resources away from primary tasks for a longer duration, even if the distraction may not need to be dealt with immediately (or may not even be possible to deal with). Nagging alarms in the background can be such a distraction, as can any type of extraneous thought or concern about things not related to the primary tasks.

Tasks that cannot be executed in their normal, practiced sequence are another source of vulnerability to prospective memory. A task may not be completed or executed in its normal sequence because information from other persons or parties is still missing, for example, or those other parties or persons are not yet present. Simply waiting for such gaps to be filled is often not practical, so the task can then be deferred while other work is continued. But deferring one task can mean that other tasks also need to be deferred. This is when mental bookkeeping becomes quite tricky. What was accomplished by whom? Normal cues for prospective memory (e.g., the completion of one task that triggers another) can fall between the cracks of practice. Another reason for the disruption of normal task sequences can be the interruptions discussed. Unanticipated new task demands might then arise. Finally, there are cases when practitioners need to remember not to execute a particular task step (because of some specific features of that patient, operation, or disease). As the step is a normal part of an oft-rehearsed sequence, it might slip in unnoticed unless active attentional resources are allocated to its prevention (or some cue is located in the environment to trigger prospective memory to that effect).

KNOWLEDGE FACTORS

Successful performance in healthcare settings obviously requires the possession of knowledge about the problem at hand. But possession is not enough. It also involves the organization and application of the knowledge, that is, the extent to which knowledge can be used flexibly but appropriately in different contexts. The possession and availability of knowledge is of course intricately interwoven with the allocation of attention, as discussed in this chapter. How do practitioners bring knowledge to bear effectively in their work and make decisions based on it? This is not just about knowledge content. Of course, the right knowledge needs to be there. It has to be complete and not erroneous or buggy. Knowledge also needs to be organized in a way that makes it possible to map knowledge onto evolving situations. Finally, knowledge, even if well organized, might still not be activated (or called to mind) in the context where it is needed because of various attentional processes (such as cognitive fixation).

BUGGY MENTAL MODELS, HEURISTICS, AND OVERSIMPLIFICATIONS

Human factors likes the idea of mental models simply because they are useful for the understanding of human performance in complex settings. This label makes a number of epistemological assumptions or assumptions about how we know what we know. The most important one is that people build up and entertain a model of how the world (or a part of it) works in their minds. And that model can be accurate or inaccurate, correct or buggy relative to the real workings. While this is a contestable assumption, human factors has produced considerable research on buggy mental models. Mental models are said to be buggy not only when they contain gaps but also when they contain incorrect assumptions or oversimplifications. Technical devices have been a topic of particular interest (Sarter & Woods, 1994) because what people know can be contrasted against the "real" functioning of the device. Practitioners may not have detailed knowledge of why a device does what it does (or even, sometimes, of what it does) and may have simplified models of input-output relations (when *I* do this, *it* does that). But even various physiological functions of patients have been the subject of this kind of research, yielding interesting data about how medical practitioners reason and remember (Feltovich, Spiro, & Coulson, 1989; Johnson et al., 1981).

An effective way of dealing with complexity is to apply heuristics or cognitive rules of thumb (recall the schemata from Chapter 2). These are approximations or simplifications that are easier to remember and quicker to apply than formal decision rules or algorithms. Heuristics help reduce the cognitive effort required to produce decisions and leave time for other work. Like all models, heuristics are necessarily distortions, misconceptions, or contractions. Heuristics that appear to work satisfactorily under some conditions can therefore produce "error" in others. Feltovich and colleagues (1989) found that medical students and practicing physicians sometimes applied heuristics that amount to oversimplification.

> Bits and pieces of knowledge, in themselves sometimes correct, sometimes partly wrong in aspects, or sometimes absent in critical places, interact with each other to create large-scale and robust misconceptions. (p. 162)

These can lead to interesting effects. For example, students or medical practitioners can sometimes see different entities as more similar than they actually are, treat dynamic phenomena as static, assume that some general principle accounts for all of the observed symptoms or, conversely, treat highly interconnected issues as separate, or treat continuous variables as discrete (Feltovich, Spiro, & Coulson, 1993).

Neither the building block approach to medical training (anatomy, physiology, etc.) nor a more case-based or problem-based approach can be entirely immune from the formation of such oversimplifications. Problem-based learning can create the impression that a particular disease is limited to the few expressions or variations that were learned in the cases. Building-block learning, on the other hand, may assume too easily that decomposed knowledge will eventually add up in the mind of people to become fuller or whole pictures of a phenomenon. Either approach, then, can produce a false sense of understanding *and* inhibit pursuit of deeper understanding. Learners tend to resist learning a more complex model once they already have an apparently useful one.

This can lead to problems with knowledge calibration—the extent to which the practitioner is aware of bugs and imperfections in his or her mental models (Woods et al., 2010). The diversity and complexity of clinical practice can mean that practitioners can remain unaware of their bugs for a long time. They may simply not encounter a situation that exposes the bugs and may not be aware that their knowledge outside of frequently used scripts and schemata is severely underdeveloped. Practitioners may in fact be able to organize their work and its encounters in ways that keep them away from these underexplored areas of knowledge. Without feedback that reveals how ill calibrated they are, practitioners will not likely invest effort in recalibration, also because this might run up against issues of identity and infallibility. Yet, expertise is not the same as having perfect knowledge. Rather, it is being well calibrated about the extent and limits of one's knowledge. Feltovich et al. (1989) explains why heuristics are so robust:

> It is easier to think that all instances of the same nominal concept ... are the same or bear considerable similarity. It is easier to represent continuities in terms of components and steps. It is easier to deal with a single principle from which an entire complex phenomenon "spins out" than to deal with numerous, more localized principles and their interactions. (p. 131)

But heuristics are not just easier. They can also be better or more effective in practice. Following more formal paths for reasoning and decision making (e.g., as suggested in some clinical guidance) can involve more work and more distractions from getting the task done. Simplified decision methods may actually produce a higher proportion of correct responses in many situations, particularly under time pressure. The point is not that either rules or heuristics are bad; the point is that both have their strengths and their limitations.

Injury to the extrahepatic bile ducts remains the most significant operative complication of cholecystectomy and is an important unsolved surgical problem, feared because it is followed by substantial morbidity, occasional death, large additional healthcare costs, and frequent litigation (Hugh & Dekker, 2008).

Laparoscopic bile duct injury is no respecter of the seniority or experience of the surgeon. The persistence of this dreadful complication, even in the hands of experienced surgeons, suggests that there are underlying systemic predispositions to the injury. Much of the published work on iatrogenic bile duct injury emphasized the need for clear identification of bile duct anatomy before dividing, clipping, or cauterizing any structure. The problem is, no surgeon would divide a structure without having identified it, and no surgeon would transect the common bile duct (CBD) during cholecystectomy knowing it was the common duct.

A human factors approach would suggest that the psychology in duct mis-identification lies in a surgeon being persuaded sufficiently, if not believing completely, that the structure being transected is the correct one, the cystic duct. Underestimation of risk and cue ambiguity are critical contributory factors to the construction of this belief, even when in hindsight it turns out to have been false. It is possible that a chronic preoperative underestimation of the risk of laparoscopic bile duct injury may contribute to the error. An underestimation of risk in complex procedures may occur on the basis of past success in avoiding the error; this is particularly likely in major bile duct injury (MBDI) because of the infrequent exposure of individual surgeons to the error. Experienced surgeons may have a sense of "this can't happen to me," unaware that past success is no guarantee of future safety.

Constructing a mental image that convinces the surgeon that the CBD or the right hepatic duct is the cystic duct amounts to the central error in many MBDIs. The possible psychological and heuristic sources of misidentification stem from difficult intraoperative decisions in cholecystectomy that have to be made from ambiguous cues. An important factor in misidentification would seem to be the absence of haptic (active touch) perception in the laparoscopic technique. Absence of haptic perception is a laparoscopy-specific problem, but data do not seem to support the assertion that misidentification was uncommon in the open cholecystectomy era. Haptic perception thus may not be critically important in duct identification, and it seems probable, therefore, that duct misidentification in both open and laparoscopic cholecystectomy is based on similar visual misinterpretation. The frequency and type of MBDIs are similar in both operative methods.

Visual perception is one form of heuristics, combining the processes of acquiring, interpreting, selecting, and organizing sensory information, especially uncertain, probabilistic information and using it as a basis for action—actions that themselves will make more information about the world available, thereby informing further action and so on (recall the perceptual cycle from Figure 3.1). What is done influences what is seen, which then helps constrain and determine what can be done. This continual interweaving of action with perception means, for example, that retracting the gallbladder in a superior rather than a lateral direction increases the risk of misidentification because it tends to align the cystic duct with the CBD.

Visual perception can be thought of as a continual mental construction that informs and is informed by interaction with the world, where "rules" or "scripts" build expectations about what we (should) see. Such constructions have a tendency to err on the side of the familiar and the expected—this is the whole point of mental and perceptual heuristics: freeing up cognitive capacity for other tasks or threats in the environment. When the subhepatic field is observed at laparoscopy, the surgeon usually matches what is seen with a learnt mental map of the "normal" biliary tree. This matching is a rapid and largely subconscious process integrated with visual perception and may sometimes be a matter of "seeing what you believe" rather than believing what you see. The familiar and the expected in this case is the "normal" pattern of the biliary tree that may be mentally superimposed on a very different ductal reality. A duct that appears to merge with the infundibulum of the gallbladder may be accepted as the cystic duct when in reality it is the CBD or the right hepatic duct.

The cyclically constructive nature of perception can also sustain this same belief during and after the procedure, even as the dissection yields new information. The usual consequence of identifying an important bile duct incorrectly as the cystic duct is that it is clipped and divided. Taking action, as Weick (1995) said, not only simplifies the problem but also implies a commitment, which in turn can produce blind spots (Weick, 1995). Once surgeons have committed themselves to a particular course of action, they will build an explanation that justifies that action and guides further action. This explanation tends to persist and is transformed into an assumption ("I am working on the correct structure") that is taken for granted during the rest of the procedure and beyond. Subsequent steps in removal of the gallbladder, for example, typically lead to an encounter with the proximal hepatic end of the divided duct (usually the common hepatic duct or the right hepatic duct), which is then divided a second time, resulting in resection of a substantial length of duct. This second cutting of the duct may not be recognized, even in the presence of unexpected intraoperative biliary leakage.

When ambiguous cues in the initial situation (and actions on it) have biased the surgeon in some direction, this promotes certain cues at the expense of others. For instance, the initial interpretation typically persists throughout the dissection and division of a mistakenly identified duct, even when extra lymphatic and vascular structures show up in close proximity, when there is nonopacification of proximal ducts on cholangiography, or when the duct cannot be fully encompassed by a 9-mm clip (which to an objective observer would seem to indicate that the duct was abnormally large for a supposed cystic duct).

Such plan continuation often persists postoperatively. In most patients with MBDI, there is delay in recognition of the injury in the postoperative period. Plan continuation may be remarkably strong and persistent, for example, causing the surgeon repeatedly to reassure juniors anxious about the patient's progress, even in the face of postoperative cues such as jaundice, biliary leakage, or signs of biliary peritonitis that may seem in retrospect obvious indicators of duct injury.

Often, none of those emergent cues is strong enough to push a surgeon off the interpretation subscribed to, or the path taken, as none of them fits the assumption that has been the basis for all action and perception to that point. Recall that this can be referred to as cognitive fixation (this and nothing else). Fixation, as in this case, is one possible side effect of a mental balancing act: Should a surgeon maintain stability of interpretation and course of action in the face of changing, contradictory, or ambiguous cues? Or, should he or she shift course of action with each newly incoming cue? Neither is desirable in most clinical situations, but ending up at one extreme (fixated on one interpretation) or the other (vacillating among multiple possible interpretations) is sometimes part of doing expert work in complex, dynamic situations. One way out is the recruitment of additional, outside expertise that has not been part of the initial formulation of the problem. But when a surgeon believes that he or she has operated on the correct structure, then there is no trigger for seeking a second opinion.

Several intraoperative heuristics have been suggested, including the retraction of the infundibulum of the gallbladder laterally to open out the triangle of Calot. An alternative technique is to retract the infundibulum medially to expose the posterior aspect of the triangle and to evaluate its relationship to Rouviere's sulcus, a landmark that indicates the plane of the CBD. A further technical strategy is for the surgeon to develop what has been described as the "critical view of safety" in which no structure is clipped or divided until the gallbladder is sufficiently free from the liver to allow visualization of just two structures entering it: the cystic artery and the cystic duct.

At least preoperatively, there is little substitute for a deliberate sense of heightened awareness of the risk. One method is for the surgeon to repeat the following as a mantra while scrubbing for a cholecystectomy: "This could be the one." Briefing and discussing the possible error with junior staff who will be present during the operation, with a review of the possible perceptual errors before, can also be a good investment in safety—the potential for a second opinion is thereby engineered beforehand into the procedure. Telling junior staff that, as all the data suggest, seniority is no safeguard against the error, which can make them feel free to speak up (see also Chapter 6).

KNOWLEDGE AND DIAGNOSIS

Kathryn Montgomery (2006) probes beyond the mere cognitive aspects of the use of heuristics and simplifications in healthcare. Discussing the simplification of clinical cause, she suggests how physicians might oversimplify, or think in linear terms about causes and effects of disease, because of their deontological commitment—their duty ethic. It is the ethics of medicine as a practice, the need to intervene in a patient's illness, that creates the pressure to reduce cause to its simplest manifestation possible. In ordinary practice, physicians will frequently maintain a simple linear model of cause despite a bewildering array of factors that they might not (yet) understand and cannot influence anyway. In the many cases for which diagnosis, therapy, and prognosis are well established, with a host of clinical and successful precedents, the whole question of cause may well slide from view. If a proximate case is known, then a plan for treating it can be created. And treatment can get reduced to a protocol that people lower down the medical competence hierarchy can execute with equal chances of success (Montgomery, 2006).

> Medicine strives for causal simplicity ... the promise of ready diagnoses with safe and efficacious treatment draws young people to medical careers—to say nothing of bringing patients to physicians. ... Thus, linear causality comes to stand for the clinical competence, the automaticity of thought, and the ready solutions to difficult problems that physicians work toward. When life or health is at risk, who does not want a what-you-see-is-what-you-get account of reality, a representation of things as they truly are, without distortion or bias? ... This ideal is regularly challenged in everyday clinical medicine. Causal simplicity is never easy for medicine to achieve because the information it needs is social and circumstantial as well as scientific. ... Medical events and conditions can be described as cellular, organic, organismic, personal, familial and cultural, and their causes can be too. What's more, cause runs both ways on the scale from cell to society since illness behavior is also social behavior, and microbial activity often depends upon it. (pp. 70–71)

This is not to say that clinicians will always revert to the simplest cause-effect relationship. In academic medicine, for example, multiplicative explanations (which fan out from symptoms to a web of possible causes, contributors, and risk factors) are both legitimate and expected and often represent an important ingredient in a physician's reputation or career advancement. Having questions unsettled is exciting, of course. It forces physicians to rethink, or renarrate the mechanisms of disease, to do more research of the literature or of colleagues' work, to question what others might previously (also) have taken for granted. Underlying these enterprises, however, is still the promise of (eventual) certainty, of settled questions about what leads to what.

Medical education has this dual task: equipping practitioners for the application of known strategies and heuristics for the pretty-much-settled questions while instilling a curiosity and skepticism that invites them to go beyond what they think they know, to become free from how they have thought before.

STRATEGIC FACTORS

In most healthcare work, contradictory goals are the rule, not the exception. Any study of human factors that does not take goal conflicts seriously does not take human work seriously. Although safety is often a (stated) priority, the systems in which people work do not exist just to be safe. Understanding human work means looking deeper into these goal interactions, these basic incompatibilities, in what people need to strive for in their work. Multiple goals are almost always active, although their contribution to people's priorities and decisions ebbs and flows as the work setting develops during a shift.

It is always informative to understand how people view and handle these contradictions and conflicts from inside their operational reality and how this contrasts with other views (e.g., of management, regulator, public) of the same activities. Operational people swiftly learn how to pursue or satisfy multiple goals simultaneously or make it look as if they do, such as "token" hand washing when going from patient to patient during busy walk arounds (Dekker & Hugh, 2010). The actual managing of goal conflicts gets pushed down into local operating units or wards; practitioners not only need to see many patients but also need to prevent infections, for example.

On a weekend in a large hospital, the anesthesiology team (four physicians, of whom three were residents) was called to perform anesthesia in one building for an in vitro fertilization, a perforated viscus, reconstruction of an artery of the leg, and an appendectomy; in another building, there was one exploratory laparotomy. None of these cases could be delayed for the regular operating room schedule. The exact sequence in which the operations were done depended on multiple factors (although on evenings and weekends, the anesthetist in charge makes these decisions). One nurse insisted that the exploratory laparotomy be done ahead of the other cases. The nurse was responsible only for that single case; operating room nurses and technicians could not leave the hospital until their case had been completed.

Meanwhile, surgeons complained about delays; their cases were becoming more urgent because of it. Preoperative preparation of some of the patients also was delayed. Available staff were able to run only two operating rooms simultaneously, so the anesthesiologist in charge attempted to overlap portions of procedures by starting one case as another was finishing.

The hospital served as a major trauma center. A team needed to be able to start a large emergency case with a notice of less than 10 minutes. In committing all residents to doing the waiting cases, the anesthesiologist in charge produced a situation in which there no longer was slack. The anesthetist in charge also was not much of a redundant resource. The role also entailed handling a variety of emergent situations in the hospital, including calls to intubate patients on the wards, serving as a backup and source of expertise for ongo-ing operations, answering requests for pain control, as well as handling new trauma cases.

The system was saturated with work. The anesthesiologist in charge resolved the goal conflict between immediate production and reserving slack for unfore-seen trauma cases by using up all productive capacity. In essence, it represented a gamble. There were no excess resources to apply to a new emergency. But this situation was so common in the hospital that it was regarded as typical rather than exceptional (Woods et al., 2010).

A huge difficulty in making goal trade-offs is uncertainty. In the example above, the anesthetist in charge could never be sure whether the gamble to deploy all resources in production would lead to trouble in case an unexpected trauma case came in. The gamble paid off this time, and this may of course be taken as a false assurance that it will pay off next time as well. What is more, management may begin to notice the productive capacity of an anesthetic team like this one and come to hold it up as the standard for others. They might never realize that important margins are being shaved off the hospital's ability to live up to its own mission and role in the community. And, the day that something goes wrong because all available resources were used up, "human error" might be conveniently blamed, and nothing is learned.

A woman was hospitalized with severe complications of an abdominal infection. A few days earlier, she had seen a physician with complaints of aches but was sent home with the message to come back in 8 days for an ultrasound scan if the problem persisted. In the meantime, her appendix burst, causing infection and requiring major surgery.

The woman's physician had been under pressure from her managed care orga-nization, with financial incentives and disincentives, to control the costs of care and avoid unnecessary procedures. The problem is that a physician might not know that a procedure is unnecessary before doing it or at least before doing part of it. Preoperative evidence may be too ambiguous. Physicians end up in difficult double binds created by the various organizational pressures (Dekker, 2006).

Multiple goals that are simultaneously active are the rule rather than the exception for virtually all safety-critical domains, especially in underresourced, overburdened organizations that provide healthcare. Practitioners in them have no choice but to cope with the presence of multiple goals—shifting between them, weighing them, choosing to pursue some rather than others, abandoning one, embracing another. Many of the goals encountered in practice are implicit and unstated, and the con-flicts between them are often left unexamined or implicit. In fact, most important goal conflicts are never made explicit. They arise from multiple irreconcilable direc-tives from different levels and sources, from subtle and tacit pressures and resource constraints, and from management or patient reactions to particular trade-offs. And, some goal conflicts are simply inherent to the nature of a domain.

In an anesthetized patient, high blood pressure pushes blood through the coronary arteries and improves oxygen supply to the heart muscle. But increased blood pressure adds to cardiac work. The appropriate blood pressure target depends in part on the practitioner's strategy, the nature of the patient, the kind of surgical procedure, and any circumstances of the case that may change (e.g., the risk of major bleeding). It may also be subject to negotiations among different people in the operating room. For example, the surgeon may like blood pressure to be kept low to limit the blood loss at the surgical site (Woods et al., 2010).

Goal conflicts also can come from the accountability context in which practitioners work. On the one hand, practitioners want to protect patient safety and avoid being sued for malpractice afterward. This maximizes the need, for example, for patient information and preoperative workup. But hospitals continually have to reduce costs and increase patient turnover, which produces pressure to admit, operate, and discharge patients on the same day.

Other factors also produce competition between different goals, for example, management policies, earlier reactions to failure (how the organization has responded to goal trade-offs that went right or wrong before), subtle coercions (do what the boss wants, not what he or she says), legal liability, regulatory and other guidelines, and economic considerations. Practitioners can also bring personal or professional interests with them (career advancement, avoiding conflicts with other groups) that enter into their negotiations among different goals. Unlike simpler situations (e.g., in a laboratory), there is typically not one best method or correct answer. Complex systems intrinsically contain many conflicts that must be resolved by practitioners at the sharp end. They face what can be called overconstrained problems, in which it is impossible to maximize the function or work product on all dimensions simultaneously.

In understanding human factors in healthcare, it can be really difficult to bring goal conflicts out in the open, precisely because practitioners are often so skilled and smooth in making them (seemingly) go away. They might make it look as if there are none. And indeed, not many of the relevant goals are written down in guidance, procedures, or job descriptions. All this makes it difficult to trace or prove the contribution of goal interactions to particular people's assessments or actions. And, it is hard for organizations, especially in highly regulated industries, to admit that these kinds of tricky goal trade-offs arise, even arise frequently.

Outsiders pay attention to practitioners' coping strategies only after a failure. Then, they might conclude that the trade-offs that were made were awkward, flawed, or vulnerable. They may even conclude that they were unethical, dangerous, or sanctionable in some way. After the fact, it is easy to say that practitioners should have prioritized one goal over another, that they should have delayed or chosen some other action that would have avoided (what people now know to be) the bad outcome. Even in such analyses, the role of goal conflicts arising from multiple, simultaneously active goals may never be noted. In fact, it is easy for organizations to produce guidance or expectations on practitioners that make certain goal conflicts worse or even introduce new ones. But denying the existence of goal conflicts does not make them disappear.

KEY POINTS

- Rather than invoking internal mechanisms of mental functioning or error generation, it is more useful to capture the cognitive functions that people must perform to handle the demands of their complex field of practice. This involves other people and technical artifacts and surrounding organizational processes and structures.
- Cognitive work is about attentional dynamics (where to focus attention, when, at the cost of what else), knowledge factors (how to learn, store, activate, and deploy knowledge), and strategic factors (how to deal with larger constraints and goals imposed by the organization in which the work is carried out).
- Healthcare presents interesting cognitive challenges to its practitioners, among them escalating problems (where the tempo, criticality, volume, and coordinative demands of tasks all increase), multitasking that denies practitioners the luxury to defer one task until another is completed, and a considerable reliance on prospective memory (remembering to do something).
- Possession of knowledge is not enough for successful cognitive performance in complex, dynamic settings. The organization and application of that knowledge in context is equally important. Buggy mental models and oversimplifications tend to make this both easier and riskier, and specific attentional mechanisms such as cognitive fixation can make the right knowledge never be called to mind, even though (in another context) the clinician can be shown to possess it.
- Contradictory goals are the rule in most healthcare work. Many goal conflicts stem from organizational or even professional incompatibilities. Practitioners not only need to see many patients but also need to prevent infections, for example. The actual managing of goal conflicts gets pushed down into local operating units or wards for individuals to sort out in the form of thousands of larger and smaller trade-offs each day. Understanding the human factors of healthcare is impossible without tracing people's goal conflicts.

REFERENCES

Cook, R. I., McDonald, J. S., & Smalhout, R. (1989). *Human error in the operating room: Identifying cognitive lock up.* Columbus, OH: The Ohio State University.

De Keyser, V., & Woods, D. D. (1990). Fixation errors: Failures to revise situation assessment in dynamic and risky systems. In A. G. Colombo & A. Saiz de Bustamante (Eds.), *System reliability assessment* (pp. 231–251). Dordrecht, Netherlands: Kluwer Academic.

Dekker, S. W. A. (2006). *The field guide to understanding human error.* Aldershot, UK: Ashgate.

Dekker, S. W. A., & Hugh, T. B. (2010). Balancing "no blame" with accountability in patient safety. *New England Journal of Medicine, 362*(3), 275.

Dismukes, R. K. (2006). *Concurrent task management and prospective memory: Pilot error as a model for the vulnerability of experts.* Paper presented at the Human Factors and Ergonomics Society 50th Annual Meeting, October 2006, San Francisco.

Feltovich, P. J., Spiro, R. J., & Coulson, R. (1989). The nature of conceptual understanding in biomedicine: The deep structure of complex ideas and the development of misconceptions. In D. Evans & V. Patel (Eds.), *Cognitive science in medicine: Biomedical modeling*. Cambridge, MA: MIT Press.

Feltovich, P. J., Spiro, R. J., & Coulson, R. (1993). Learning, teaching and testing for complex conceptual understanding. In N. Fredericksen, R. Mislevy, & I. Bejar (Eds.), *Test theory for a new generation of tests*. Hillsdale, NJ: Erlbaum.

Hancock, P. A., & Weaver, J. L. (2005). On time distortion under stress. *Theoretical Issues in Ergonomics Science, 6*(2), 193–211.

Hugh, T. B., & Dekker, S. W. A. (2008). Laparoscopic bile duct injury: Understanding the psychology and heuristics of the error. *ANZ Journal of Surgery, 78*(12), 1109–1114.

Johnson, P. E., Duran, A. S., Hassebrock, F., Moller, J., Prietula, M., Feltovich, P. J., et al. (1981). Expertise and error in diagnostic reasoning. *Cognitive Science, 5*(3), 235–283.

Loukopoulos, L. D., Dismukes, K., & Barshi, I. (2009). *The multitasking myth: Handling complexity in real-world operations*. Farnham, UK: Ashgate.

Mintzberg, H. (1979). *The structuring of organizations: A synthesis of the research*. Englewood Cliffs, NJ: Prentice-Hall.

Montgomery, K. (2006). *How doctors think: Clinical judgment and the practice of medicine*. Oxford, UK: Oxford University Press.

Neisser, U. (1976). *Cognition and reality: Principles and implications of cognitive psychology*. San Francisco: Freeman.

Orasanu, J. M., & Martin, L. (1998). Errors in aviation decision making: A factor in accidents and incidents. In *Human Error, Safety and Systems Development Workshop (HESSD) 1998*. Retrieved February 2008, from *http://www.dcs.gla.ac.uk/~johnson/papers/seattle_hessd/judithlynnep*

Sarter, N. B., & Woods, D. D. (1994). Pilot interaction with cockpit automation II: An experimental study of pilots' model and awareness of the flight management system. *International Journal of Aviation Psychology, 4*(1), 1–29.

Weick, K. E. (1995). *Sensemaking in organizations*. Thousand Oaks, CA: Sage.

Weick, K. E., & Sutcliffe, K. M. (2007). *Managing the unexpected: Resilient performance in an age of uncertainty* (2nd ed.). San Francisco: Jossey-Bass.

Woods, D. D. (1988). Coping with complexity: The psychology of human behavior in complex systems. In L. P. Goodstein, H. B. Andersen, & S. E. Olsen (Eds.), *Tasks, errors, and mental models*. New York: Taylor and Francis.

Woods, D. D., Dekker, S. W. A., Cook, R. I., Johannesen, L. J., & Sarter, N. B. (2010). *Behind human error*. Aldershot, UK: Ashgate.

Woods, D. D., & Patterson, E. S. (2000). How unexpected events produce an escalation of cognitive and coordinate demands. In P. A. Hancock & P. Desmond (Eds.), *Stress, workload and fatigue*. Mahwah, NJ: Erlbaum.

Xiao, Y., Milgram, P., & Doyle, J. (1997). Capturing and modeling planning expertise in anesthesiology: Results of a field study. In C. Zsambok & G. Klein (Eds.), *Naturalistic decision making* (pp. 197–205). Mahwah, NJ: Erlbaum.

4 New Technology, Automation, and Patient Safety

Chapter 3 showed how people's goals, knowledge, and attention are created, influenced, and enacted in the environment in which they work. Remember that this is the original human factors commitment: understanding the relationship between the success and failure of human work on the one hand and the settings in which that work is carried out on the other. That relationship is not haphazard or random; there are systematic connections between human performance and features of people's tools, tasks, and organizational settings. Influencing the way in which work, equipment, or an organization is designed will have consequences for the expression of people's error and expertise.

If there is one factor that has changed the work setting more than any others in healthcare, it is technology. Technology has dramatically changed the expression of both error and expertise. It has opened new pathways to failure, even while closing other ones. The changes created by new technology are not letting up—in fact, there are plenty of signs of accelerating developments. Healthcare in many countries is embracing electronic patient records, for example, and information and computer technology are becoming ever more deeply woven into almost all the stages of medicine, from diagnosis to intervention. Rather than getting lost in the details of specific medical technologies, however, this chapter discusses generic patterns of human–technology interaction that many people in healthcare may recognize in their workplaces. It then discusses ways in which human factors methods might help create a better fit between technology and the tasks it is supposed to support. Examples are used to illustrate the various ideas.

THE SUBSTITUTION MYTH

The original idea behind many technological interventions is that technology can do a task better, faster, or cheaper than human beings. Technology is seen as one way out of the "human error problem" (see Chapter 2). If technology does the work, then humans cannot make errors in doing that work. Or if there is technology that checks the human, then errors can be caught before they have any effects. This is the idea of substitution. Technology substitutes for human work.

But the idea of substitution is a myth. The problem is that the introduction of new technology creates *new* human work. And by creating new human work, technology introduces new opportunities to do that work well or less well. With new technology, people will have to spend time remembering input modes or understanding display readings, for example. This creates new opportunities for error and new pathways to failure.

> Simple substitution of human work is impossible because technology always changes the very tasks it was designed to support. Take the task of setting an intravenous line. The insertion of an increasingly intelligent infusion pump between the fluid bag and the patient (a pump that may even be connected to a remote database of drugs and prescriptions) does not remove the need for human work. But it does change the nature of that work. Now there are buttons to push, displays to read, patient codes to remember. All of that is new work with new opportunities to do it well or incorrectly. And there is, of course, the basic task of finding a pump. This may not always be easy in an overloaded emergency department (ED) on a Saturday night.

Substituting one medium (e.g., a computer database) for another (e.g., paper) also quickly becomes complex. Take electronic patient records. Converting to electronic databases that are accessible independent from the physical location of a patient record makes a whole host of assumptions about what makes human work and interaction easier across a hospital (or even across different care facilities). These assumptions may or may not be correct, but reading a patient record or adding something to it is never going to be a simple conversion from paper to computer screen. The possibilities that electronic data offer are just too great for that (with everything from automatic reminders, drug prescription suggestions or limitations, easily accessible treatment plans, to real-time collaborative diagnosis only mouse clicks away). The opportunities for getting it wrong are also enormous, of course. Systems like this easily become cumbersome, difficult to work with, vulnerable to lock up or break down, or open to hacking.

When things go wrong with technology in hospital settings, it is easy to invoke versions of the substitution myth. People should have followed the instructions more closely, or they should not have interfered with the technology. Conversely, they should have intervened more aggressively when the technology was not doing things well. These are all versions of the substitution myth. They are based on the premise that technology replaces human work or simply supports existing human work, a simple reciprocation of tasks that are swapped between human and machine. But technology creates all kinds of hard-to-foresee reverberations that not many investigations are capable of pulling out of the rubble. One important feature is that human adaptations are necessary to make the technology work in actual practice:

> The complexity and complications of advanced information technology are not widely appreciated. After accidents, the reconstruction of events tends to make it seem that the human performance was at fault, while the technology performed well. Closer examination, however, may demonstrate that the human performance was awkward because the humans involved were the adaptable elements of the system. To make the technology work requires a variety of adaptations and workarounds in order to get the job done. When the limits of adaptation are reached and failure occurs, human performance is evaluated and found wanting. After accidents, the adaptations are found to be

vulnerable and the workarounds are treated as violations. But these findings are more reflections of the naïveté of the finders than a meaningful assessment of the system itself. (Cook, Nemeth, & Dekker, 2008, p. 10)

Adaptations are a symptom of the mismatch between expectations of how technology will work on the one hand and what happens when it hits a field of ongoing practice on the other. As has been observed in other settings, new technology often does not always work as designed, and takes a lot of human ingenuity to function in practice: "Almost without exception, technology did not meet the goal of unencumbering the personnel operating the equipment. ... Systems often required exceptional human expertise, commitment, and endurance" (Cordesman & Wagner, 1996, p. 25).

Perversely, such difficulties in getting technology to perform correctly in practice can become a resource for identity formation or affirmation among physicians and other healthcare workers. Part of what might denote an exceptional doctor is his or her ability to get an obstinate and abstruse piece of technology to work with exceptional expertise, commitment, and endurance. This can inspire awe among more junior staff, and even patients. Such an interaction between ill-matched technology and physician identity is obviously not conducive to changes in manufacturers' design and evaluation practices. In extreme cases, particular difficult-to-work technologies can become the dominion of specialized groups in a hospital, who may exploit the user unfriendliness of the technology for their own influence and indispensability among other practitioners.

Collectively, these effects point to the hugely transformative effects of technology across healthcare settings. New technology is not just about doing one practitioner's task faster, better, or cheaper. New technology overhauls human roles and human work. It retools human relationships. It affects organizations' expectations on practitioners' performance and financial results and reshapes power distributions among practitioner groups. It remakes the face of practitioners' expertise and the expression of error and can alter the very notion of practitioner identity. Effects such as these are often underestimated by those responsible for acquisition and are hard to foresee by those who develop and design the technology.

THE TRANSFORMATIVE EFFECTS OF TECHNOLOGY

Technology qualitatively changes the setup and nature of human work. What that means is that the introduction of a new piece of equipment is not just an addition or a replacement to an existing way of working. It is not just the manipulation of a single variable while keeping all others constant. Technological change transforms the workplace, the whole system made up of people and technological artifacts (Woods, Dekker, Cook, Johannesen, & Sarter, 2010).

Computerization has tremendously advanced our ability to collect, transmit, and transform data. It is now easy to bombard users with computer-processed data, especially when anomalies occur. Of course, the human ability to digest and interpret all these data (particularly because time for this, or for learning this, is generally not available in hospitals) has failed to keep pace with our abilities to generate and manipulate increasing amounts of data. The problem that occurs is often known as *data overload* (Woods, Patterson, & Roth, 2002). Take electronic patient records, for instance. They can reproduce all of a patient's laboratory results but may then show all of them without any selectivity. Human effort is required to wade through all the results and pick out the data that are informative relative to the patient's current problems (Hartzband & Groopman, 2008).

Interface technology has created the possibility to concentrate this ever-expanding field of data into one physical platform. Users often have to access the data by looking at or manipulating a single display. Behind that display, however, lies a potentially vast array of data, not all of which are important or relevant to the task at hand. Finding what you need may be difficult. Extracting meaning from what you see on the screen may be even more difficult. This is known as the *keyhole effect*, as if the practitioner is peeping through a keyhole to find the relevant piece of data in a room full of possibilities behind the closed door (Woods, 1995). Of course, the freedom of software allows increased degrees of freedom for data handling and presentation.

Designing this well, however, has not proven easy. One physician at a major cancer center that recently switched to electronic patient records commented how chart review during rounds has become virtually worthless. He would search in vain through meaningless repetitions in multiple notes, hunting for a single line that represented a new development. The time he wasted, and the low success rate in finding the relevant data, caused him to revert to his index card-based system for patient notes (Hartzband & Groopman, 2008).

Technology (and software in particular) has also expanded the range of subtasks and cognitive activities that can be automated. The idea is often that automated resources can offload practitioner tasks. Automated monitoring of numerous patient parameters is just one of those possibilities. Computerized systems that assess or diagnose the situation at hand, alert practitioners to various concerns, and advise practitioners on possible responses are quite another. But automation, just like any other kind of technology, creates new human work. This may not even be work at which humans are best (Bainbridge, 1987) and can create the potential for what has become known as *automation surprises* (Sarter, Woods, & Billings, 1997).

Computerization and automation integrate or couple more closely different parts of the system. This is true for electronic patient records, for example. But it may also apply to the range of apparatuses used in operative anesthesia. Increasing the coupling within a system has many effects on the kinds of cognitive demands that practitioners face. With tighter coupling, actions produce more side effects, and fault diagnosis becomes more difficult. A fault is more likely to produce a cascade of disturbances that spreads throughout the monitored process. Making sense of what exactly is going wrong can be difficult in those situations, no matter what kinds of displays, alarms, or warnings are available to the practitioner. In fact, those displays and warnings can exacerbate confusion (Perrow, 1984; Snook, 2000; Woods & Patterson, 2000). As the following case shows, the deeply woven couplings and interconnections between various computer systems can make the reason for failures entirely opaque for users at the surface. Automation that was intended to provide extra layers of safety, redundancy, quantity control, and double checking paradoxically ended up creating a befuddling and dangerous situation (Perry, Wears, & Cook, 2005).

Perry et al. (2005) described how a critically ill patient came to a busy ED in a large urban area. The resuscitation nurse went to obtain medications from an automated dispensing unit (ADU), part of a computer-based dispensing system in use throughout the hospital. He found an uninformative error message on the computer screen ("Printer not available") and an unresponsive keyboard. The system did not respond to any commands and would not dispense the required medications.

The ED nurse abandoned efforts to get the ADU to work and asked the unit clerk to notify the main pharmacy that the ADU was "down," and emergency medications were needed. He asked another nurse to try other ADUs in the ED. Other ED staff became aware of the problem and joined in the search for the sought-after drugs. Some were discovered on top of another ADU in the ED, waiting to be returned to stock.

Anticipating the patient's deterioration, the ED physicians opened the resuscitation cart ("crash cart") and prepared to intubate the patient using the small amount of medications and equipment stored there.

A pharmacist came to the ED and examined the unresponsive ADU. He decided not to use the bypass facility for downtime access because neither the drawers nor the bins were labeled with the names of the medications they contained, and this information could not be obtained from a nonfunctioning unit. Instead, he arranged for the pharmacy staff to use runners to bring medications from the main pharmacy (one floor below) to the ED. The patient eventually received the requested medications, and her condition improved. She survived and was later discharged from the hospital (Perry et al., 2005).

The automated dispensing system had been installed a few years before to improve inventory tracking and reduce errors and pilferage. Except for some resuscitation drugs stored in crash carts, hospital medications were dispensed via this system, with 40 ADUs linked to two centrally located computers through the hospital information system (HIS). For safety, the ADUs were programmed to deny access to a drug unless there was a current valid pharmacist-approved order for it in the HIS. This feature, however, had not been activated in the ED because of the time constraints associated with ED drug orders and delivery.

Two weeks before the incident, a major HIS software upgrade had been disrupted by a sudden, unexpected hardware failure that resulted in the complete loss of all HIS functions. In response, operators in the pharmacy disabled the safety interlock feature that required order checking before dispensing medications so that nursing staff on the wards could obtain drugs. As the HIS came back online, the pharmacy operators enabled this feature to restore normal operations. The HIS, however, repeatedly crashed during this process and required pharmacy staff to enable and disable the safety interlock feature various times. The computer then executed these commands for each item in the inventory of each dispensing unit. This created a huge storm of messages to and from the dispensing units that slowed the system response such that individual units appeared to be unresponsive to user keyboard commands, similar to denial-of-service attacks on the Internet (Perry et al., 2005).

As the case suggests, practitioners tend to learn the strengths and pitfalls of new technology quickly, and they will try to prevent failing in their work by adapting or tailoring the technology to their needs and understandings and by developing and modifying failure-sensitive strategies. They might, for example, put reminders on

Post-it notes beside a display or develop their own heuristics for achieving particular settings in an automated device. Or they might open a crash cart if there is no other way to access critical medications. Such strategies, however, imply extra memory burdens or task load for the human and may be brittle in the face of subtle changes in the context in which the technology is used. A heuristic that works brilliantly with a piece of technology for adult patients, for instance, may break down or become dangerous when the same heuristic is applied to pediatric patients.

NEW TECHNOLOGY BEYOND TASK-SPECIFIC DEVICES

The example case also reveals aspects particular to healthcare that might make it extra vulnerable to failures of this kind compared with other safety-critical industries (Perry et al., 2005). The healthcare field is relatively new to the implementation of large, integrative, and complex computer technology beyond highly task-specific devices such as infusion pumps or imaging devices. The narrow focus of new technology on limited and specific tasks has now been supplemented by computer systems that are more broadly aimed at addressing organizational problems. These may include billing, inventory control, accounting, and patient records. Significantly, these systems are only secondarily directed at supporting clinical work, if at all (Nemeth, Nunnally, O'Connor, Klock, & Cook, 2005). There are not many people in hospitals who are able to meaningfully assess the impact of such technological changes on people's daily work and how it might affect error, expertise, and ultimately patient safety (Cook, Potter, Woods, & McDonald, 1991; Wears & Perry, 2002; Xiao, Milgram, & Doyle, 1997).

The other interesting thing revealed by the incident in the example is the deeply conflicted nature of healthcare as an enterprise. These computer systems advance organizational goals, but an already beleaguered group of practitioners then needs to engage with and service the technology, providing it with inputs and interpreting its outputs or sluicing them forward to other professionals. Local benefits for the practitioners who may need to do this are often a bit elusive. Instead, existing tensions between different professional groups (who may compete for scarce resources or are grudgingly dependent on each other's performance or deliveries) can be amplified or exacerbated (Perry et al., 2005).

The substitution myth suggests that the advantages of new technology are basically quantitative. Hopes of quantitative, measurable improvements typically accompany the introduction of systemwide technology. Management might think it can measure the effect of the technology along single dimensions, all of which are better for the bottom line, reputation, or waiting lists of a hospital. New technology means reduced cost for a procedure (at least in the long run), less education or training of the practitioners operating it (reducing cost there), fewer human errors and better discovery and mitigation of the errors that are still made, better measures of the administration of controlled drugs, more efficient administering of a test, or quicker results.

Particular technologies may or may not live up to such promised measures, but that hardly matters from a human factors perspective. Rather than providing quantifiable changes, the reverberations of new technology are qualitative, and that is where the potential for success or failure really lies. The research showed clearly, however, that new technology leads to the emergence of new human roles and the exacerbation of existing or creation of new organizational conflicts and tensions. Technology changes what is routine about people's work and what is exceptional; it affects what people need or no longer need from other practitioners in the system and when they need it. It changes the kinds of errors that people can make and the kinds of expected or required interactions with other professionals or departments. The consequences of errors might be mitigated by the technology, or they might be amplified. Paths to failure and paths to success become different from previous ones, and the relationship to patients may be irrevocably altered.

Electronic patient records, to cite that example again, have introduced the risk of a physician spending much more time watching a computer screen than watching the patient and asking prefabricated questions (prompted by the electronic system) that restrict the range of diagnostic exploration. Interfaces for the electronic patient record system have a direct influence on how, and to what extent, this patient–doctor relationship becomes transformed. If boxes need clicking and particular fields need filling in, then there may be little time for engaging in a narrative-based, open-ended dialogue with the patient, and clinical thinking may be blunted into preformed templates (Hartzband & Groopman, 2008; Montgomery, 2006).

The remainder of this chapter looks at two human factors aspects of new technology in more detail: data overload and automation surprises. It concludes with considering what might need to be done to oversee and test medical technology for its usability and sensitivity to human factors.

DATA OVERLOAD

A generic problem with new technology is linked to its capability for gathering, storing, and presenting ever-growing amounts of data. For the users of the technology, data overload may be the resulting experience. This is actually a paradoxical problem. The plentiful availability of data should in principle be a benefit to practitioners' problem solving, but for people to find meaning in these potential floods of data is not easy. The ability to make sense of data, in other words, has not kept pace with the ability to generate and display it. Each new generation of technology seems to exacerbate the data overload problem rather than solve it.

The data overload problem can be characterized in at least three ways, each of which implies particular recipes for how to address it (Woods et al., 2002). The first is to see data overload as a clutter problem: There is too much data on a screen; there is simply too much stuff. In this case, data overload can be addressed, in principle, by reducing the number of data units displayed in space or over time. One problem with that, of course, is that things may be removed that actually are critical for a practitioner to see in any given situation. Interesting changes and developments may be hidden from the practitioner underneath a deceptively simple display, and waiting for the practitioner to ask for more data creates new problems. As alluded to in the discussion of multitasking in Chapter 3, practitioners may not be aware of the need even to look at other data without cues in their environment. Also, asking for more data might sound simpler than it is, given the interface and interaction properties of the machine. How to bring new data forward on a small screen can demand significant human memory capacity and may not always be successful.

Another characterization of the data overload problem is that of a workload bottleneck. In this case, data overload is seen as the problem of having too much to do, to respond to, or to check in the time available. The way to address the problem is by using automation and other technologies to perform activities for the user or putatively to cooperate with the user during these activities. As will be discussed in this chapter, automating parts of people's tasks so that they are not overloaded with data introduces new challenges, such as automation surprises.

A third way to look at data overload is as a problem of finding the significance, or meaning, of data when the practitioner does not know a priori which data will be informative. This characterization requires a completely different approach, and it might well be the most difficult one. What is interesting to a practitioner, of course, depends on the context. The meaning, or even relevance, of data hinges in large part on what else is happening. Information, in that sense, is a relationship between data and between the data and the practitioner's expectations, intentions, and interests (Woods et al., 2002). Meaning lies also in contrasts, in departures from some reference or expected course, and human attention has evolved to quickly notice and deal with surprises, novelty, and contrast. Solutions to the data overload problem, then, will foremost have to help practitioners put data into context. This partly shifts the burden of finding out about meaningfulness back to the designer, who needs to be aware of the messy details of practice in that setting—the ebbs and flows of work, the multiple other tasks that need attending, and the roles and responsibilities distributed across a team of human problem solvers. Rather than organizing displays around data, data can be organized around meaningful questions and typical problems of clinical practice in that setting (Flach, 2000).

AUTOMATION SURPRISES

Automation is often developed in the hope of increasing the precision and economy of operations while reducing operator workload and training requirements (Sarter et al., 1997). The idea is that it is possible to create an autonomous (sub)system that requires little, if any, human involvement. This supposedly reduces or even eliminates the opportunity for human error; it may also help reduce the data overload problem. In these hopes, we also see the substitution myth at work: Automation can replace human work without any larger impact on the system in which that action or task occurs, except on some measures of its output.

But again, it is not that simple. Automation does not typically reduce the workload associated with a task across the board. Rather, it redistributes workload. In practice, it will reduce human work during traditionally low-workload periods in practice but increases workload when the practitioner is traditionally busy already. Interaction burdens tend to accrue during precisely those times when practitioners can least afford it, which has led to the characterization of "clumsy" automation: a redistribution of workload over time rather than an overall decrease or increase (Wiener, 1989). Automation creates new communication and coordination demands. During busy times, automation technology might need human inputs, such as setting target parameters, and it might require close monitoring to ensure that commands have been received and are carried out as intended. These human tasks may not be supported well; it can be difficult to make correct inputs or to discover what is going on, or there may be too many alarms or indications to be meaningful any longer (Moray, 1984; Wiener, 1988).

Automation distributes workload unevenly not only over time but also sometimes between operators who have to work around the technology as a team. Some members may be a lot busier during certain times, while others have little to do (Sarter et al., 1997). The very nature of workload may also change, again producing a qualitative rather than a quantitative shift. For example, practitioner work may shift from active control of multiple parameters to supervisory control of an automated system doing the same things. This imposes new attentional demands, and it requires that the operator knows more about his or her systems to be able to understand, predict, and intervene in their behavior (Bainbridge, 1987).

This in turn creates new demands on human expertise that did not exist previously. Training on new technology in healthcare is often nominal. In fact, some technology in healthcare is introduced without any formal training or checks of personnel. Even if there is training, the complexity of many modern systems cannot be fully covered with the time and resources available. The consequence is that practitioners typically learn a subset of techniques, input–output relationships, or recipes with which they can make the system work under routine conditions. There is often no systematic support to help practitioners discover and correct bugs in their model of the automation. Practitioners may figure out how to work the system but have no idea how the system works (Sarter et al., 1997).

Miscalibrated mental models are the result, and these can be persistent (see also Chapter 3). Areas of incomplete or inaccurate knowledge of device operations can remain hidden from practitioners because they have the capability to work around these areas by limiting themselves to a few well-practiced and well-understood methods. Situations that force practitioners into areas where their knowledge is limited and miscalibrated may arise infrequently. Another factor that contributes to poor knowledge calibration about device operation is ineffective feedback on the state and behavior of automated systems. It may simply be difficult to see what the machine is actually doing, let alone understand why (Billings, 1997; Norman, 1990).

After initial system setup, automated systems can be capable of carrying out long sequences of action without any further human commands. This autonomy can produce situations in which state or mode changes occur and the human practitioner is not aware of them. This can be related to programmed delays between user input and system execution, or the system itself shifts its behavior based on changes in the parameters it monitors. This can lead to what has been called automation surprises. The automated system does something that the human had not expected, often because of autonomous mode changes. Automation surprises are more likely when there is a call for practitioners to direct their attention elsewhere and it may be busier or more critical than normal. Interestingly, the first signs of anomalies in automation behavior are often not noticed in the machine (another testimony to often-underdeveloped feedback) but in the monitored process itself (e.g., the patient's condition) (Sarter et al., 1997).

The human factors literature is well developed when it comes to problems with designing feedback (Billings, 1997; Sanders & McCormick, 1993). Some things that should be avoided, for example, include nuisance communication, such as voice alerts that talk too much in the wrong situations, excessive false alarms, or distracting indications when tasks that are more serious are being handled. Automated systems, however, must give clear feedback to the human practitioner when they are moving to the limits of their authority and when their ability to act on a deteriorating situation becomes compromised. There have been situations in which feedback to that effect was so poor that humans were unable to meaningfully intervene when the capability of the automation was finally exhausted (Dekker & Woods, 1999). These are just a few of the many things that need to be considered from a human factors perspective—an exhaustive treatment of which falls outside the scope of this book.

EVALUATING AND TESTING MEDICAL TECHNOLOGY

It could be argued that there is something vaguely akin to a cottage industry in the development and design of particularly task-specific medical technology. Medical technological innovation in clinical niches has not infrequently been the outgrowth of the personal engagement of a particularly inspired practitioner. The advantage of such sources of technological innovation is that there is both the commitment and the expert knowledge of what it takes to do a certain task better or make it safer. The disadvantage is the undercapitalization of design and development and the fact that devices may be introduced that seem brilliant to their maker, but that succeed in befuddling many other practitioners who are expected to work with them.

Markets for such technology are not generally large, and they are highly specialized. This even goes for a software suite that might run an electronic patient record system. With thousands of hospitals in the United States alone, developing software for such an application might seem like a golden opportunity, but it pales when compared to the sale of software licenses in, for example, office applications. As a result, investing many resources up front in human factors research or usability testing may not be high on anybody's agenda. This may even apply to larger healthcare information technology projects:

> The oversight and evaluation of large healthcare information technology systems is in disarray. These systems are being developed and implemented throughout healthcare with little thought being given to their potential for harm or the difficulties associated with their use. The enthusiasm for new technology as a means to save money and rationalize care is not matched by the performance of these systems. (Cook et al., 2008, p. 11)

One problem is the slim regulation of medical devices that requires a priori evaluation or assessment of technology. The definition of medical device applied by the U.S. Food and Drug Administration (FDA), for example, is purposely narrow. This may encourage innovation, but it puts broader, hospital-wide systems such as electronic patient safety records or ADUs outside the regulator's reach. Even devices such as infusion pumps have largely escaped meaningful regulatory scrutiny, introducing a host of risks to patient safety as a result.

> Infusion pumps often provide critical fluids to patients, so failures or wrong volumes or dosages can have significant effects (Nemeth et al., 2005). An estimated 2 million infusion pumps are used in hospital and clinical settings in the United States alone, and hundreds of thousands more are used by patients in their homes. The pumps use a variety of designs to intravenously deliver food, fluids, and drugs such as pain medications, insulin, and cancer treatments.
>
> In 2010, the U.S. FDA made its first moves toward stricter regulation of infusion pumps, issuing a proposal for new guidelines. These guidelines advise manufacturers of the devices on which they will have to run more in-depth clinical trials before the FDA clears new pumps (Phillips, 2010).
>
> With an initial focus on insulin pumps, the FDA convened a panel to review the risks posed by the pumps after 18 recalls were issued for the devices over a 5-year period. The agency's General Hospital and Personal Use Devices Panel discussed the findings of an agency report that looked into possible health risks related to insulin pump failures. There had been almost 17,000 adverse events reported for insulin pumps from 2006 to 2008. Of those, the FDA said between 300 and 700 deaths and 12,000 injuries occurred, possibly because of problems caused by pump failures.
>
> Pump manufacturers responded by saying that most problems occur when a nurse or other health care worker enters the wrong data accidentally. FDA officials said, however, that based on their review of pump complaints, they thought many deaths and injuries related to the devices were less the result of user error than of product design and engineering (Phillips, 2010). For example, agency officials said that some pumps were prone to key bounce, a problem in which defective software interprets a single keystroke as two separate presses of that key. For example, instead of dispensing 2 units of a drug, a pump would dispense 22 units.
>
> Thus the FDA moved to establish additional premarket requirements for infusion pumps, in part through issuance of a new draft guidance and letter to infusion pump manufacturers. The draft guidance proposed that manufacturers be required to provide additional design and engineering information to the agency during premarket review of the devices. Producers would be required to provide additional data to support the procedures they used to determine the effectiveness and safety of their devices. In addition, companies would have to conduct limited clinical trials to ensure that their pumps were not susceptible to misuse or had design elements that could create errors. The FDA also sent a letter to infusion pump manufacturers informing them that they may need to conduct additional risk assessments to support clearance of new or modified pumps. Manufacturers could voluntarily submit the software code of the devices for analysis prior to the review by the FDA of an application.

Of course, providing a larger role for regulatory pressure assumes that the regulator would know what to examine. The clinical trials suggested by the FDA for infusion pumps, for example, are not the same as conducting usability studies or doing a thorough human factors design assessment. Compared to other safety-critical worlds, healthcare has historically been a relatively insular, isolated field of practice. Many of the problems are solved or addressed (or left to fester) in-house, and there is a general lack of awareness of the knowledge and experience with design and human factors engineering that might usefully be applied to healthcare systems (Xiao et al., 1997). This means that even if people in healthcare inquire about the potential adverse effects of a new device or computer system, they may not be aware of available methods and expertise. They might, as usual, convene a group of their own practitioners, who may have good intentions but are not well versed in questions of usability design, human factors, or risk assessment (Perry et al., 2005).

Of course, new technology brings new capabilities for practice. But other typical results also occur, not least because of accompanying organizational changes. New capabilities offered by technology tend to increase demands on humans, for example, by expecting a higher tempo or volume of operations. It also creates new complexities. The ADU example showed increased coupling across previously loosely connected parts of the system, for instance. The clumsy use of technology can create surprises and open new pathways to failure. Of course, adaptations by practitioners normally hide the complexities from designers and managers. Any failures that occur, then, might be ascribed to an occasional human error.

Given the increased pace and volume of technological changes and possibilities in healthcare, success will come to those projects that can predict the transformations, changing roles, organizational reverberations, and the kinds of adaptations practitioners will likely develop to cope with new complexities. Better still, they should also foresee situations that will challenge these strategies and adaptations to anticipate future errors and failure. It is crucial to use these predictions early in the design process to avoid the negative unintended side effects of technology change (Woods & Dekker, 2000).

THE ENVISIONED WORLD PROBLEM

The ultimate purpose of new technology, and the role of human factors in its development, is to stimulate design and innovation to discover new ways to use technological possibilities to enhance human performance. Interestingly, these possibilities seem less constrained by questions of feasibility (computing power is abundantly available, for example) and more *how* to use the possibilities skillfully to meet operational and other goals—without creating undesirable side effects. One role for human factors is to help determine what will be useful in future work.

Assessing the impact of design changes on envisioned or future practice in healthcare is of course difficult. There are traditional human factors tools for the verification, validation, and (lately) certification of isolated aspects of new systems. Yet, such piecemeal "V&V" may amount to an oversimplification fallacy—that understanding multifactor, interconnected processes, let alone anticipating longer-term evolutions of tool tailoring and task transformation, can derive from the momentary assessment of the state of a few independent things, objects, artifacts, or subprocesses. As discussed, design changes and their effects are never insular; they always interact with other systems and with psychological and social aspects of practice. They become tangled up in organizational, procedural changes as well, affecting areas not predicted by designers. Can designers actually anticipate the full range of such changes? There is always a gap between the envisioned impact of a new development and the actual impact and reverberations of that change. That gap may be so large or so daunting that one conclusion could be that a new piece of technology can ultimately be validated only in full-fledged operation. By then, of course, any necessary changes could have become unaffordable.

The problem created by, for example, the ADUs is one of envisioned worlds: large, comprehensive, interconnected operational systems that have yet to be fielded (Dekker & Woods, 1999). If you would ask around a hospital, it is easy to obtain multiple versions of how proposed changes will affect the interdependent fields of practice in the future. Different stakeholders have different perspectives on the impact of new objects on the nature of practice—theirs or other people's. The downside of such plurality is a kind of parochialism: People mistake their partial, narrow view for the dominant view of the future of practice and are unaware of the plurality of views across stakeholders. This may depend critically on who is in charge of selecting and implementing such new technology and who is involved and consulted in the process. The upside of plurality is the triangulation that is possible when the multiple views are brought together. In examining the relationships, overlaps, contradictions, and gaps across multiple perspectives, hospitals can at least grapple with the inherent uncertainty of looking into future practice (Dekker, Mooij, & Woods, 2002).

Another feature of the envisioned world problem is underspecification. As a hypothesis about the impact of new objects on the nature of practice, each envisioned concept is of necessity vague on many aspects. The upside of underspecification is the freedom to explore new possibilities and new ways to relax and recombine constraints. This can lead to innovation and improvement. The downside of underspecification is the risk of remaining trapped in a disconnected, shallow, unrealistic view of practice. When the view of practice is disconnected from the pressures, challenges, and constraints operating in that world, it will be distorted and miss how strategies of practice are adapted to constraints and pressures common to the field.

In the development of envisioned worlds, validity (or perhaps better, authenticity) derives from (1) the extent to which problems to be solved in the test situation represent the vulnerabilities and challenges that exist in the target world and (2) the extent to which real problem-solving expertise is brought to bear by the study participants. Studies into human performance in envisioned worlds can rate high on both of these measures, for example, by thinking up future incidents that practitioners will have to be able to handle (Dekker & Woods, 1999). In other words, these studies must investigate real practitioners caught up in solving real domain problems. In this way, system development can be steered toward more fruitful, cooperative solutions.

Envisioning a coevolving, dynamic, and future process of change and adaptation is highly uncertain. People who push the technology (and this may be anybody from practitioner groups to hospital management) can easily miscalibrate and become overconfident that if the systems can be realized, the predicted consequences and only the predicted consequences will occur. There is a way to hedge against this. People's views of the future should not be seen as partially finished prototypes but rather as tentative hypotheses, as conjectures of what might (not) be useful. As such, these people need to remain open to revision and subject their hypotheses to empirical jeopardy. All who envision, design, and develop new technology for healthcare are susceptible to the fragility of their own envisioned stories. All should speculate about the impact of the object to be created as a source of manifold changes in a field of practice. Devoting attention to the messy details of practice can generate the kind of knowledge that is both detailed and early enough to have a doable and affordable impact on the direction of technology change.

KEY POINTS

- Technology is often seen as one way out of the "human error problem." The original idea behind many technological interventions is that technology can do a task better, faster, or cheaper than human beings. If technology does the work, then humans cannot make errors in doing that work. Or, if there is technology that checks the human, then errors may be caught before they have any effects.
- The idea of substitution is a myth. The introduction of new technology creates *new* human work, requiring new expertise and introducing new opportunities for error. Data overload and automation surprises are only two of the obvious symptoms. Technology is not just an addition or a replacement to an existing way of working, the manipulation of a single variable while keeping all others constant. Technological change transforms the workplace, including many human-to-human relationships.
- Computerization and automation also integrate or couple more closely different parts of the system, which creates new vulnerabilities and pathways to failure. Such systems are often created to advance organizational goals, expecting practitioners to engage with and service the technology. Local benefits may remain elusive. Instead, it can amplify existing tensions between different professional groups (who may compete for scarce resources or are dependent on each other's performance or deliveries).
- While much technology in healthcare is introduced without any formal training or checking of personnel, practitioners tend to learn the strengths and pitfalls of new technology quickly, and they will try to prevent failing in their work by adapting or tailoring the technology to their needs and understandings and by developing and modifying failure-sensitive strategies. These adaptations may be so smooth that designers or managers are not even aware of the cognitive energy that is needed to put into them.
- Perversely, in fact, getting user-unfriendly technology to work in practice can be a resource for identity formation or affirmation among physicians and other healthcare workers. Getting an obstinate or abstruse piece of technology to work with exceptional expertise, commitment, and endurance can inspire awe among more junior staff and patients.

REFERENCES

Bainbridge, L. (1987). Ironies of automation. In J. Rasmussen, K. Duncan, & J. Leplat (Eds.), *New technology and human error* (pp. 271–283). Chichester, UK: Wiley.

Billings, C. E. (1997). *Aviation automation: The search for a human-centered approach.* Mahwah, NJ: Erlbaum.

Cook, R. I., Nemeth, C., & Dekker, S. W. A. (2008). What went wrong at the Beatson Oncology Centre? In E. Hollnagel, C. P. Nemeth, & S. W. A. Dekker (Eds.), *Resilience engineering perspectives: Remaining sensitive to the possibility of failure.* Aldershot, UK: Ashgate.

Cook, R. I., Potter, S. S., Woods, D. D., & McDonald, J. S. (1991). Evaluating the human engineering of microprocessor-controlled operating room devices. *Journal of Clinical Monitoring, 7*(3), 217–226.

Cordesman, A. H., & Wagner, A. R. (1996). *The lessons of modern war, Vol. 4: The Gulf War.* Boulder, CO: Westview Press.

Dekker, S. W. A., Mooij, M., & Woods, D. D. (2002). Envisioned practice, enhanced performance: The riddle of future (ATM) systems. *International Journal of Applied Aviation Studies, 2*(1), 23–32.

Dekker, S. W. A., & Woods, D. D. (1999). To intervene or not to intervene: The dilemma of management by exception. *Cognition, Technology and Work, 1*(2), 86–96.

Flach, J. M. (2000). Discovering situated meaning: An ecological approach to task analysis. In J. M. Schraagen, S. F. Chipman, & V. L. Shalin (Eds.), *Cognitive task analysis* (pp. 87–100). Mahwah, NJ: Erlbaum.

Hartzband, P., & Groopman, J. (2008). Off the record—Avoiding the pitfalls of going electronic. *New England Journal of Medicine, 358*(16), 1656–1658.

Montgomery, K. (2006). *How doctors think: Clinical judgment and the practice of medicine.* Oxford, UK: Oxford University Press.

Moray, N. (1984). Attention to dynamic visual displays in man-machine systems. In R. Parasuraman & D. R. Davies (Eds.), *Varieties of attention.* New York: Academic Press.

Nemeth, C., Nunnally, M., O'Connor, M., Klock, P. A., & Cook, R. (2005). Getting to the point: Developing IT for the sharp end of healthcare. *Journal of Biomedical Informatics, 38*(1), 18–25.

Norman, D. A. (1990). The problem with automation: Inappropriate feedback and interaction, not over-automation. *Philosophical Transactions of the Royal Society of London Series B–Biological Sciences, 327*(1241), 585–593.

Perrow, C. (1984). *Normal accidents: Living with high-risk technologies.* New York: Basic Books.

Perry, S. J., Wears, R. L., & Cook, R. I. (2005). The role of automation in complex system failures. *Journal of Patient Safety, 1*(1), 56–61.

Phillips, T. (2010). *General hospital and personal use devices panel meeting.* Washington, DC: Food and Drug Administration.

Sanders, M. S., & McCormick, E. J. (1993). *Human factors in engineering and design* (7th ed.). New York: McGraw-Hill.

Sarter, N. B., Woods, D. D., & Billings, C. (1997). Automation surprises. In G. Salvendy (Ed.), *Handbook of human factors/ergonomics.* New York: Wiley.

Snook, S. A. (2000). *Friendly fire: The accidental shootdown of U.S. Black Hawks over northern Iraq.* Princeton, NJ: Princeton University Press.

Wears, R. L., & Perry, S. J. (2002). Human factors and ergonomics in the emergency department. *Annals of Emergency Medicine, 40*(2), 206–212.

Wiener, E. L. (1988). Cockpit automation. In E. L. Wiener & D. C. Nagel (Eds.), *Human factors in aviation* (pp. 433–462). San Diego, CA: Academic Press.

Wiener, E. L. (1989). *Human factors of advanced technology ("glass cockpit") transport aircraft* (No. 117528). Moffett Field, CA: NASA Ames Research Center.

Woods, D. D. (1995). Towards a theoretical base for representation design in the computer medium: Ecological perception and aiding human cognition. In J. Flach, P. Hancock, J. Caird, & K. Vicente (Eds.), *An ecological approach to human-machine systems I: A global perspective.* Hillsdale, NJ: Erlbaum.

Woods, D. D., & Dekker, S. W. A. (2000). Anticipating the effects of technological change: A new era of dynamics for human factors. *Theoretical Issues in Ergnomics Science, 1*(3), 272–282.

Woods, D. D., Dekker, S. W. A., Cook, R. I., Johannesen, L. J., & Sarter, N. B. (2010). *Behind human error.* Aldershot, UK: Ashgate.

Woods, D. D., & Patterson, E. S. (2000). How unexpected events produce an escalation of cognitive and coordinate demands. In P. A. Hancock & P. Desmond (Eds.), *Stress, workload and fatigue.* Mahwah, NJ: Erlbaum.

Woods, D. D., Patterson, E. S., & Roth, E. M. (2002). Can we ever escape from data overload? A cognitive systems diagnosis. *Cognition, Technology and Work, 4*(1), 22–36.

Xiao, Y., Milgram, P., & Doyle, J. (1997). Capturing and modeling planning expertise in anesthesiology: Results of a field study. In C. Zsambok & G. Klein (Eds.), *Naturalistic decision making* (pp. 197–205). Mahwah, NJ: Erlbaum.

5 Safety Culture and Organizational Risk

The human factors view of patient safety says that the major source of risk lies not with individual caregivers but with the system surrounding those caregivers: the organization, administration, design, resourcing, and technology of healthcare. The previous chapters have laid out how individual practitioners' cognitive and coordinative processes of care delivery are enabled and constrained by this larger system. The reconciliation of goal conflicts, the management of fatigue and production pressures, the provision and maintenance of knowledge and skills, the direction of attention in noisy, multitasking environments—all these processes are in large part produced and influenced by how care is organized, administered, and technically supported.

By implication, then, this is where risk to patients brews and grows and where risk should be most effectively recognized, managed, and contained. You could say that a complex system like a hospital has a sharp end and a blunt end. At the sharp end, practitioners interact directly with the patient or with other related hazardous processes. It is at the sharp end that care eventually flows through the hands of the caregiving individuals. At the blunt end, regulators, administrators, economic policy makers, and technology suppliers control the resources, constraints, and multiple incentives and demands that sharp-end practitioners must integrate and balance. Practitioners at the sharp end continuously adapt their strategies to cope with the complexities of the processes they monitor, manage, and control and to make do with the resources and constraints from the blunt end of the system.

Research findings about complex system failure turn the "blunt end" of a hospital into an area not only of interest for study but also of potential intervention to increase safety. Changing things there will change things in the environment in which front-end practitioners work, and it will thus affect their chances of success and failure.

In this chapter, we consider how theorizing around the blunt end has evolved since the 1970s. It presents various models and theories that attempt to explain how risk builds up and is controlled (or not) through the administration and management of safety-critical processes. Distinctions between these different models are always to some extent arbitrary, but here they are divided up as follows:

- First, the chapter takes on models that equate organizational risk with energy to be contained. This includes man-made disaster theory and the so-called Swiss cheese model. The pedigree of those models is not hard to guess: industrial accidents where huge releases of energy were not contained or controlled. What is their solution to risk? It is to build stronger barriers or put in more barriers between the object to be protected and where the energy is located. Note how adverse events, in these models, can be prevented by stopping linear sequences of events.

- Then the chapter considers a model that sees organizational risk as a structural property of complex systems. No matter how much critical reflection an organization engages in, this model says, risk will not go away because it is structurally embedded in the way the system interacts and is coupled. The only way to reduce risk is to reduce complexity.
- The chapter moves on to models that see risk as the gradual organizational acceptance of the abnormal. The solution to risk is to ensure that the organization continually reflects critically on and challenges its own definition of "normal" operations, and finds ways to prioritize chronic safety concerns over acute production pressures.
- Finally, the chapter discusses models that see risk as a managerial or control problem. The potential for failure builds, according to these models, because deviations from the original design assumptions of the system become increasingly rationalized and accepted. Their solution to risk lies in a repertoire of organizational activities and managerial commitments and countermeasures.

None of these models is necessarily right or wrong. They are, after all, just models. Their conception of the world can be read into any setting, and for some settings some models may be more effective than others. What matters is the diversity of ideas about what risk is, how it builds at the organization's blunt end and how it can be controlled. This diversity means that there are a number of different ways in which those in healthcare can conceptualize its risk and engage in attempts to control it. For some situations, one particular model may be exceptionally effective (e.g., putting in an extra barrier, which reduces risk), but for another problem this may not work (extra barriers will create complexity, which increases risk). It is probably a good idea to consider an area of risk from all the angles that these models suggest and then assess what might work best. Risk can be seen as something that is managed by putting in extra barriers, by questioning the notion of normal operations, by reducing complexity, or by enhancing the control system and managerial or leadership commitments surrounding it. Taking a combination of all these visions of risk, and their concomitant countermeasures, could be the safest option in many cases.

SAFETY CULTURE AND DRIFTING INTO FAILURE

A safety culture is a culture that allows the boss to hear bad news. This presents two problems, a relatively easy one and a really hard one. The relatively easy one relates to creating an organization in which news actually gets to the boss. Practitioners need to feel relevant and empowered to share information with the boss. Structures and processes for making the news flow need to exist, and management needs to show commitment to such news and doing something about it. In some cases, the only way to get bad news to the boss is actually not to report it to the boss at all but to use channels outside line management (e.g., confidential reporting to a safety/quality staff), which in turn will get the news to the boss. These topics are discussed in the next chapter. It is a hard problem, for sure, but not as hard as the really hard problem.

The hard problem is to decide what is bad news. Most models of risk in this chapter recognize that an entire operation or organization can negotiate or shift its idea of what is normative and thus negotiate or shift what counts as bad news. Bad news, in other words, is a critical category for these models and for making the management of risk work. But the whole idea of "bad news" (i.e., an error or near miss) is subject to negotiation and interpretation. Medical culture can see adverse events as signs of individual incompetence or simply as bad luck. Neither of those interpretations is likely to push a practitioner to report. The bad news, even if seen as bad, is not worth reporting because bad luck is immune to organizational intervention, and incompetence is something that can only be solved by adding skills or removing the practitioner.

But consider the following: One in seven hospitals in the United States alone reported nursing vacancy rates of over 20%. This is actually not unique to the United States, and it is not going to get better any time soon. One-third of the nurses are over age 50, leaving a shortfall of up to a million nurses once the baby boom generation will have to receive care in hospitals (Wachter, 2008). Understaffed wards, in other words, are normal. In fact, understaffing by 2020 may be so acute that the 20% vacancy rate could be looked back on as good or safe. The work got done, did it not? So, where really is the evidence of "understaffing"? And who sets those staff norms against which hospitals are supposedly 20% short? Social systems have all kinds of ways of adapting their expectations and language around such fundamental constraints and make the unusual and deviant look normal and compliant.

Any failures that do occur can be rationalized away by the organization or its management (they are the result of "nurses who make mistakes," for example; the system just needs to get rid of those nurses—see Chapter 1). Any other costs in terms of practitioner burnout or low job satisfaction are considered collateral damage or "externalities." These collect elsewhere and are not really counted against the productive capacity of the system. Yet there are data that show that adverse event rates increase with higher patient–nurse ratios. Surgical patients, according to one study, had a 31% higher chance of dying in the hospital when the average nurse cared for more than seven patients. And, for every additional patient, mortality rose 7% (Aiken, Clarke, Sloane, Sochalski, & Silber, 2002). Similar effects can be found for interns/residents or junior doctors. One study showed that as the number of patient admissions per on-call resident went up, so did patient length of stay, costs, and mortality rate (Ong, Bostrom, Vidyarthi, McCulloch, & Auerbach, 2007).

This sets out how safety culture and organizational risk are related to adverse events. Iatrogenic harm is not caused by a coincidence of independent failures and human errors. Rather, it is the result of a systematic migration of organizational behavior under the pressure of operating in an underresourced, aggressive environment (Rasmussen, 1997). Operational success in such resource-constrained environments implies exploitation of the benefit from operating at the fringes of usual, accepted practice. But the more success is attained at (or even beyond) those fringes, the more normal practicing there will become, the more accepted, or even the more *expected.* If others can do it (like performing 18 bypass operations in a single day or working 120 hours per week), then what is anybody else's excuse for not doing so? The routine violation of the 80-hour resident workweek limit set by the U.S. Accreditation Council for Graduate Medical Education (or ACGME) is an example of constant normative drift, as is the notion of nurse understaffing. A 20% understaffing level can become normal, tolerated, workable, as can a 90-hour workweek, even after the regulatory intervention of the ACGME. Bad news ceases to be bad news. It becomes normal. And once 20% understaffing or a 90-hour workweek have become the norm, then 21% understaffing should also be tolerable or a 95-hour workweek. After all, it is only a small step. There should be no large effect in patient mortality and so on.

This is the major residual risk in complex systems: a slow but steady drift into failure as the definition of risk and normal operations is constantly renegotiated under the pressure to save resources and accomplish multiple goals simultaneously. This, then, is the yardstick against which models of safety culture and organizational risk should be assessed. How can they meaningfully deal with the risk of drifting into failure? Are they targeted and sensitive enough to capture how practitioners and groups in a healthcare organization are constantly negotiating (mostly tacitly) what is bad news and what is not?

RISK AS ENERGY TO BE CONTAINED

In 1966, a portion of a coal mine tip (unusable material dug up in the process of mining coal) on a mountainside near Aberfan, South Wales, slid down into the village and engulfed its school. There were 144 people killed, including 116 children. The postdisaster inquiry waded into a morass of commissions, bodies, agencies, and parties responsible (or not) for the various aspects of running a coal mine, including the National Coal Board, the National Union of Mineworkers, the local borough council, the local planning committee, the engineering office of the borough, the Commission on Safety in Mines, Her Majesty's Inspector of Mines and Quarries, and a local member of Parliament. Collectively, there was a belief that tips posed no danger, that mining was dangerous for other reasons (which were generally believed to be well controlled through these multifarious administrative and regulatory arrangements). Danger in a tip was allowed to build up, while the belief remained in place that everything was okay—until the tip slid into the village.

MAN-MADE DISASTER THEORY

A researcher named Barry Turner (1978) found that the inquiry reports of this and other disasters contained a wealth of data about the administrative shortcomings and failures that precipitated those accidents. Yet, there was no coherent framework, no theory, that could do something sensible with all these data. His man-made disasters theory changed that. Disasters, it said, are incubated. Prior to the disaster, there is a period in which the potential for disaster builds. This period contains unnoticed or disregarded events that are at odds with the taken-for-granted beliefs about hazards and the norms for their avoidance. Turner considered managerial and administrative processes as the most promising for understanding this discrepancy between a buildup of risk and the sustained belief that it is under control. The space that gradually opened up between preserved beliefs and growing risk was filled with human agency—perceptions, assessments, decisions, actions. Man-made disaster theory thus shifted the focus from engineering calculations of reliability to the softer, social side of failure. It shifted focus from structures to processes, spread out over time, people, and groups and organizations, all ironically more or less given the task of preventing adverse events from happening.

The disaster incubation period was of course socially, psychologically, and organizationally the most fascinating. This is the phase when drift happens. It is characterized by the "accumulation of an unnoticed set of events which are at odds with the accepted beliefs about hazards and the norms for their avoidance" (Pidgeon & O'Leary, 2000). This discrepancy between the way the world is thought to operate and the way it really is, breeds slowly but surely, and

> rarely develops instantaneously. Instead, there is an accumulation over a period of time of a number of events which are at odds with the picture of the world and its hazards represented by existing norms and beliefs. Within this "incubation period" a chain of discrepant events, or several chains of discrepant events, develop and accumulate unnoticed. (p. 72)

Those beliefs collapse after a failure has become apparent, although of course much psychological and political energy can go into attempts to preserve those beliefs. This is the problem of converting a fundamental surprise into a local one: Instead of overhauling the beliefs in the safety of the system and the ways to ensure that safety, the disaster will be seen as a local glitch, due to some technical component failure or human error. Of course this is much easier, cheaper, and more convenient than reconstructing beliefs about safety and risk (and having to make concomitant system investments). Disasters, for Turner (1978), were primarily sociological phenomena. They represent a disruption in how people believe their system operates, a breakdown of their own norms about risk and how to manage it.

A failure comes as a fundamental surprise, as a shock to the image that the organization has of itself, of its risks and how to contain them. The developing vulnerability has long been concealed by the belief in the organization that it has risk under control, a belief that it is entitled to according to its own model of risk and the imperfect organizational-cognitive processes that help keep it alive. Indeed, how that belief is kept alive and is reproduced fascinated Turner (1978). He found how people's erroneous assumptions helped explain why events went unnoticed or misunderstood. This, he said, had to do with rigidities of human belief and perception and with the tendency to disregard complaints or warning signals from outsiders (who are not treated as credible or privy to inside knowledge about the system), as well as a reluctance on the part of decision makers at many levels to fear the worst outcome. He identified what he called "decoy phenomena," distractions away from the major hazard. For example, at Aberfan, residents mistakenly believed that the danger from tips was associated with the tipping of very fine waste rather than more coarse surplus material. When it was agreed that fine waste would not be dumped on tips in the same way, they withdrew some of their protests.

Turner (1978) also saw managerial and administrative difficulties in handling information in complex situations that blurred signal with noise. There were failures to comply with discredited or out-of-date regulations, and these passed unnoticed because of a cultural lag in what was accepted as normal. In other words, "bad news" was well hidden, even if there was no obvious intention by any participant in the system to do so. Then there was the "strangers and sites" problem: people (strangers) entering areas they officially should not because of a lack of mandate or knowledge, and sites being used for purposes that were not their original intention. These all amounted to the errors and communication difficulties that Turner saw as critical for creating the discrepancy in which failure was incubated.

In addition to the idea of incubation and the growth of a gap between how the world operates and how people think it operates, there was another important innovation in man-made disaster theory. This was the idea that a successful system produces failure as a normal, systematic by-product of its creation of success. As said, an operationally successful system operates at the fringes of normal, accepted practice, which is precisely why it can be successful. The potential for failure does not build up at random, as if it were some abnormal, irrational growth alongside and independent from normal organizational processes. On the contrary, man-made disaster theory suggested how the potential for an accident accumulates precisely because the organization, and how it is configured in a wider administrative and political environment, is able to make opportunistic, nonrandom use of organized systems of production. Those same systems are responsible for the organization's success. A new multifunction linear accelerator that can work in both x-ray mode and electron mode is a great example (a case example is given further in this chapter). It saves space and acquisition and training costs. Patient treatment rates can go up, as the machine can be made to operate around the clock and in whatever desired mode. In the same vein, dumping mining refuse on a tip just outside the coal mine, which necessarily abuts the village so that people can live close to their work, helps make the whole system work. It keeps the enterprise economically sustainable without a lot of unnecessary transport of people and stuff. The whole arrangement seems unproblematic until an adverse event reveals that it is not.

Food Poisoning as Man-Made Disaster

Take the processing of food in one location. This offers greater product control and reliability (and thus, according to current ideas about food safety, better and more stringent inspection and uniform hygiene standards). Such centralized processing, however, also allows effective, fast, and widespread distribution of unknowingly contaminated food to many people at the same time precisely thanks to the existing systems of production. This happened during the outbreaks of food poisoning from the *Escherichia coli* bacterium in Scotland during the 1990s (Pidgeon & O'Leary, 2000).

Man-made disaster theory suggested how system vulnerability arises from the unintended and complex interactions between seemingly normal organizational, managerial, and administrative features. The sources of failure must be sought in the normal processes and relationships that make up organizational life, not in the occasional malfunctioning of individual components.

The models here have adopted their ideas about risk from industrial safety improvements of the first half of the twentieth century. Consistent with this lineage, they suggest we think of risk in terms of energy, such as a dangerous buildup of energy, unintended transfers, or uncontrolled releases of energy (Rosness, Guttormsen, Steiro, Tinmannsvik, & Herrera, 2004). The overdosing of potent medication could be conceptualized this way. This risk needs to be contained, and the most popular way is through a system of barriers: multiple layers whose function it is to stop or inhibit propagations of dangerous and unintended energy transfers. This separates the object to be protected from the source of hazard by a series of defenses (which is also a basic notion in the latent failure model). Other countermeasures include preventing or improving the recognition of the gradual buildup of dangerous energy (something that inspired man-made disaster theory); reducing the amount of energy (e.g., reducing vehicle speeds or the available dosage of a particular drug in its packaging); or preventing the uncontrolled release of energy or safely distributing its release.

Medication errors have indeed been a popular target of energy-based interventions. Unit dosing was developed in the 1960s; drugs are delivered to hospital pharmacies and wards in ready-to-use doses instead of large bottles of pills or intravenous medications. This practice has now become almost ubiquitous in Western hospitals. In addition, the use of automatic dispensing machines, which are increasingly computerized and linked to centralized inventory systems, is spreading (Wachter, 2008). These are all interventions that target the unintended releases of large doses of energy by reducing the amounts that are available at any one time. The removal of intravenous potassium from wards is an example for which the potential energy (which could be unintentionally released and stop a heart, for example) is taken out of the system altogether. Finally, bedside double checks of dosages and delivery rates are a final barrier before the energy is released into the patient. These checks should preferably be done by somebody who was not involved in prescribing or preparing the drug.

A problem is that barrier-based interventions only conceptualize risk one way (energy to be contained) and thus miss other ways to look at the problem. For example, the medication error problem can also be seen as one of complexity (see Chapter 8), and some barriers increase that complexity rather than create simplicity. Ultimately, the medication error problem can be seen as related to the growing importance of the pharmacological model of disease management. Through pharmacological interventions, many diseases have become manageable, if not curable. With extended life expectancies and lifelong drug dependencies, there is a lot of money to be made. Of course, this is related to the expectation on the part of individual patients that their problem will be taken care of (the succor that medicine should provide). But perhaps a part of that succor is given at the cost of not developing alternatives such as hyperbaric medicine, lifestyle or environmental changes to precursors of the disease, epigenetic research or other nonpharmacological interventions, in some instances even surgery. The risk of medication errors, in this case, cannot be locked away behind barriers of unit dosing, bar coding, naming, packaging, or double checking. Rather, it requires taking on the cultural expectations and assumptions as well as the economic configurations of modern healthcare. Such wider reflections are of course completely out of reach for barrier models.

Radiation Energy

An example of a risk that can be conceptualized as uncontained releases of energy is radiation overdosing. The Therac-25 radiation therapy machine (technically called a medical linear accelerator [linacs]) became infamous because of its ability to release huge amounts of uncontrolled energy. Like other linacs, it accelerates electrons to create high-energy beams that can destroy tumors with minimal impact on the surrounding healthy tissue (Leveson & Turner, 1993). Radiation treatment machines were in enormous demand in hospitals throughout North America, and the Therac equipment, principally controlled by software, was widely considered the best in a growing field.

After a number of suspicious cases of overdosing by Therac-25 machines, a physicist and technician became fascinated by a "malfunction 54" message that had flashed on the screen during a treatment (the patient fell into a coma and died 3 weeks later). They typed and retyped the prescription into the computer console, determined to re-create malfunction 54. They went to the bottom of the screen and then moved the cursor up to change the treatment mode from x-ray mode to electron mode, repeatedly, for hours. Finally, they nailed the malfunction.

What made the difference was the speed with which the instructions were entered. The computer would not accept new information on a particular phase of treatment (in this case, changing the x-ray mode to electron mode) if the technician made the changes within 8 seconds after reaching the end of the prescription data. That is what malfunction 54 meant (but did not say). If the changes were made so soon, all the new screen data would look correct to the technician. But, in the computer, the software would already have encoded the old information.

That meant the beam on the Therac-25 was set for the much stronger dose needed for an x-ray beam while the turntable was in the electron position. The computer software included no check to verify that various parts of the prescription data agreed with one another (N. G. Leveson & Turner, 1993).

The conceptualization of risk as energy to be contained or managed has its roots in efforts to understand and control the physical nature of accidents. This also points to the limits of such a conceptualization. It is not necessarily well suited to explain the organizational and sociotechnical factors behind system breakdown or equipped with a language that can meaningfully handle processes of gradual adaptation, risk management, and human decision making. The central analogy used for understanding how systems work in these models is a technical system. And the chief strategy for understanding how these work and fail has always been reductionism. That means dismantling the system and looking at the parts that make up the whole. This approach assumes that we can derive the macro properties of a system (e.g., safety) as a straightforward combination or aggregation of the performance of the lower-order components or subsystems that constitute it. Indeed, the assumption is that safety can be increased by guaranteeing the reliability of the individual system components and the layers of defense against component failure so that accidents will not occur. This assumption also lies in part behind the conflation of quality and safety in healthcare (see Chapter 8).

How to Avoid Man-Made Disasters

Man-made disaster theory argues that (Pidgeon & O'Leary, 2000) "despite the best intentions of all involved, the objective of safely operating technological systems could be subverted by some very familiar and 'normal' processes of organizational life" (p. 16). Such "subversion" occurs through usual organizational phenomena such as information not being fully appreciated, information not correctly assembled, or information conflicting with prior understandings of risk. Turner noted that people were prone to discount, neglect, or not take into discussion relevant information, even when available, if it mismatched prior information, rules, or values of the organization.

The problem is that this does not really explain how or why people who manage a ward, service, or hospital are unable to "fully" appreciate available information despite the good intentions of all involved. In the absence of such an explanation, the only prescription for them is to try a little harder and to realize that safety should be their main concern, to try to imagine how risk builds up and travels through their organization. Indeed, man-made disaster theory offers that a "good" safety culture both reflects and is promoted by at least the following four features (Pidgeon & O'Leary, 2000):

- Senior management commitment to safety
- Shared care and concern for hazards and a willingness to learn and understand how they have an impact on people
- Realistic and flexible norms and rules about hazards
- Continual reflection on practice through monitoring, analysis, and feedback systems

Through these four aspects, an organization can continuously monitor and revise what constitutes bad news for them. Also, it allows them to improve and enhance the way in which they deal with bad news. Not all organizations, or even units or services within one hospital organization, are equal in this. In other work on how organizations deal with bad news, Ron Westrum (1993) identified three types of organizational culture that shape the way people respond to evidence of problems:

- *Pathological culture.* Suppresses warnings and minority opinions; responsibility is avoided, and new ideas are actively discouraged. Bearers of bad news are "shot"; failures are punished or covered up.

- *Bureaucratic culture.* Information is acknowledged but not handled. Responsibility is compartmentalized. Messengers are typically ignored because new ideas are seen as problematic. People are not encouraged to participate in improvement efforts.
- *Generative culture.* This type is able to make use of information, observations, or ideas wherever they exist in the system, without regard to the location or status of the person or group having such information, observation, or ideas. Whistleblowers and other messengers are trained, encouraged, and rewarded.

Westrum's (1993) generative culture points to some of the activities that offer an organization maximum access to bad news. Lower-level employees are empowered to come with bad news; they are even encouraged and rewarded to do so. As high-reliability theory has also shown, such empowerment of lower-ranking organizational members taps into a huge well of local, situated knowledge about what is risky and what is not. Not surprisingly, the empowerment of nurses is one of the main platforms for the improvement of patient safety in Peter Pronovost's account (Pronovost & Vohr, 2010).

THE SWISS CHEESE MODEL

The Swiss cheese model (also known as the latent failure model or defenses in depth model) emerged in the late 1980s. It preserves the basic features of the risk-as-energy model. Defenses need to be put in place to separate the object to be protected from the hazard. They are measures or mechanisms that protect against hazards or lessen the consequences of malfunctions or erroneous actions. These defenses come in a variety of forms. They can be engineered (hard) or human (soft); they can consist of interlocks, procedures, double checks, actual physical barriers, or even a line of tape on the floor of the ward (which separates an area with a particular antiseptic routine from other areas, for example). According to Reason (1990), the "best chance of minimizing accidents is by identifying and correcting these delayed action failures (latent failures) before they combine with local triggers to breach or circumvent the system's defenses." This is consistent with ideas about barriers and the containment of energy or the prevention of uncontrolled release of energy.

None of these layers of defense is perfect. They have "holes" in them. An interlock can be bypassed, a procedure can be ignored, a safety valve can begin to leak. An organizational layer of defense, for example, involves such processes as goal setting, organizing, communicating, managing, designing, building, operating, and maintaining. All of these processes are fallible and produce the latent failures that reside in the system. This is not normally a problem, but when combined with other factors, they can contribute to an accident sequence. Indeed, according to the latent failure model, accidents happen when all of the layers are penetrated (when all their imperfections or holes line up). Incidents, in contrast, happen when the accident progression is stopped by a layer of defense somewhere along the way. The Swiss cheese model got its name from the image of multiple layers of defense with holes in them. Only a particular relationship between those holes, however (when they all "line up"), will allow hazard to reach the object to be protected. The Swiss cheese model relies in this on the sequential or linear progression of failures idea that became popular in the 1930s, particularly in industrial safety applications. There, adverse outcomes were viewed as the conclusion of a sequence of events. It was a simple, linear way of conceptualizing how events interact to produce a bad outcome.

According to the sequence-of-events idea, events preceding the accident happen linearly, in a fixed order, and the accident itself is the last event in the sequence. It also has been known as the domino model for its depiction of an accident as the end-point in a string of falling dominoes (Hollnagel, 2004). Consistent with the idea of a linear chain of events is the notion of a root cause—a trigger at the beginning of the chain that sets everything in motion (the first domino that falls and then, one by one, the rest). The sequence-of-events idea has been pervasive. It forms the basic premise in many risk analysis methods and tools such as fault-tree analysis, probabilistic risk assessment, critical path models, and more. Some sequence-of-events models depict multiple parallel or converging sequences (like in some root cause analyses, RCAs). This tries to capture the greater complexity of the precursors to an accident. What gets pointed to as the root cause in any of this, however, is of course arbitrary. Because even that root cause is itself the effect of things (other causes) that lie yet further or deeper.

An important point was brought home by the Swiss cheese model. Multiple failures are all necessary and only jointly sufficient to let the hazard reach the place it should not be. The model thus helped direct the focus away from frontline operators and toward the upstream conditions that influenced and constrained their work. As Reason put it in 1990:

> Rather than being the main instigators of an accident, operators tend to be the inheri-tors of system defects created by poor design, incorrect installation, faulty mainte-nance and bad management decisions. Their part is usually that of adding the final garnish to a lethal brew whose ingredients have already been long in the cooking. (p. 173)

This invokes the idea by man-made disaster theory of the incubation of factors prior to the adverse event itself. Reason (1990) refers to hidden or resident "patho-gens" in an explicit analogy to viral processes in medicine. Resident pathogens, or "latent failures," refer to errors or failures in a system that produce a negative effect but whose consequences are not revealed or activated until some other enabling condition is met. Latent failures are decisions or other issues whose adverse conse-quences may lie dormant within the system for a long time, only becoming evident when they combine with other factors to breach the defenses of the system (Reason, 1990). Some of the factors that serve as "triggers" may be errors on the sharp end, technical faults, or atypical system states. Latent failures are associated with manag-ers, designers, maintainers, or regulators—people who are generally far removed in time and space from handling incidents and accidents.

Organizational blunt-end factors shape practitioner cognition and create the potential for erroneous actions and assessments. The clumsy use of technology (see Chapter 4) can be construed as one type of latent failure. This type of latent failure arises in the design organization. It predictably leads to certain kinds of unsafe acts on the part of practitioners at the sharp end and contributes to the evolution of incidents toward disaster. Task and environmental conditions are typically thought of as "performance-shaping factors." These also can be seen as latent failures (e.g., nurse staffing shortage or resident fatigue). The best chance of minimizing accidents is by learning how to detect and appreciate the significance of latent failures before they combine with other contributors to produce disaster (Reason, 1990).

Of course, with the inclusive and broad definitions of this model, anything can potentially be seen as a latent failure, which makes targeting them difficult. Also, the depiction of an organization as a static set of layers presents problems. It does not explain how such latent failures come into being or how they actually combine with other factors to push a failure trajectory along to a bad outcome. The model does not tell how layers of defense are gradually eroded, for example, under the pressures of production and resource limitations and overconfidence based on successful past outcomes.

Yet, the Swiss cheese model sets error in organizational context. The concept of latent failures highlights the importance of organizational factors. It reminds people how practitioners at the sharp end are constrained and influenced by the larger system in which their work is embedded. Or, as Vaughan (1996, p. 114) put it, "Individual behavior cannot be understood without taking into account the organizational and environmental context of that behavior." The model shows clearly that it takes more than an error by a frontline practitioner to produce an adverse event. Interventions that only target the last line of defense leave people there at the mercy of everything that flows down to them through upstream layers. This reminder of context has been a valuable contribution to the discussion about patient safety.

RISK AS COMPLEXITY

Barriers are designed to stop a linear progression of errors and failures before they contribute to adverse events. The principle of barriers is to separate objects to be protected from sources of hazard. Multiple redundant mechanisms, safety systems, and guidelines, policies, and procedures are meant to keep systems from failing in ways that produce bad outcomes. The results of combined operational and engineering measures make these systems relatively safe from single-point failures; that is, they are protected against the failure of a single component or procedure directly leading to a bad outcome.

But, the paradox is that such barriers and redundancy can actually add complexity, increase opacity, and increase risk. This happens for at least three reasons. The first is that redundant systems are often less independent than we think. Having somebody else double check a medication preparation, for example, might seem like a good barrier, one that is independent from the original calculations. But, the preparation and its calculations may be shown to the person who is double checking *by* the person who made the preparation, *together* with the original prescription. Any error potential in the original prescription also resonates in the double check (so there is no independence), and, by explaining what was done, the person who prepared the drug may put the person who is double checking on a garden path to accepting any erroneous conclusions (there are examples of this; Dekker, 2007). Social interaction to do the double check, in other words, can destroy the independence of the double check.

The second reason is that adding redundancy can make the system more opaque and harder to understand. Individual or component failures may be less visible and remain unfixed, and latent problems can accumulate over time (Sagan, 1993). And, when even small things start going wrong, it becomes exceptionally difficult to get off an accelerating pathway to system breakdown. Barriers, whether in the form of physical obstructions or policies, procedures, protocols, and other redundant mechanisms, typically add to the complexity of a system. With the introduction of each new part, procedure, protocol, or other type of defense layer, there is an explosion of new relationships (*between* parts and layers and components) that spreads out through the system.

Think of the introduction of a new procedure to double check something that was implicated as a broken part or process in some previous adverse event. The new procedure relates to the old procedure and its remnants in people's memories and rehearsed action sequences. It relates to people who need to carry it out in context, to people who have to train the new procedure, to the regulator who may need to approve it. It may take time and attention away from other tasks, which in turn can create a host of reverberations throughout the system. As shown in Chapter 4, the introduction of automation or computer technology as a layer of redundancy has produced the same sorts of ironies. It creates new human work and introduces new pathways to breakdown. In fact, the explosive growth of software has added greatly to the complexity of systems. With software, the possible final states of a system can become mind boggling.

Third, adding redundancy can make the system appear safer. Organizations often take advantage of such putative improvements by pushing the system to accomplish more, taking it to new production levels. Under resource pressure, any safety benefits of change can quickly be sucked into increased productivity, which pushes the system back to the edge of the performance envelope. Most benefits of change, in other words, come in the form of increased productivity and efficiency and not in the form of a more resilient, robust, and therefore safer system. Hirschorn spoke of this observation as the law of stretched systems (Woods & Hollnagel, 2006): "We are talking about a law of systems development, which is every system operates, always at its capacity. As soon as there is some improvement, some new technology, we stretch it."

The need to make these systems reliable through various kinds of redundancy, then, also makes them complex, introduces new pathways to failure, and potentially creates ill-calibrated ideas about risk. Barriers (safety systems, redundancies, double checks) can *add* risk to the system. This insight has grown into what is known as normal accident theory (Perrow, 1984). Normal accident theory holds that adverse events are not the result of a few or more component failures (human and machine). Rather, adverse events involve the unanticipated interaction of a multitude of events in a complex system—events and interactions whose combinatorial explosion can quickly outwit people's best efforts at predicting and mitigating bad outcomes.

Adverse events, says normal accident theory, are the structural and virtually inevitable product of systems that are both interactively complex and tightly coupled. Interactive complexity and coupling were proposed as two different dimensions. Normal accident theory predicts that the more tightly coupled and complex a system is, the more prone it is to suffering a "normal" accident. Or, to put it differently, the tighter the coupling and the greater the interactive complexity, the more normal it will be to have adverse events occur. They are the structural by-product of the system, and the only way to reduce their potential is to reduce complexity: to reduce couplings and interactions (Woods, Patterson, & Cook, 2005).

Interactive complexity refers to interactions between parts of the system that are nonlinear, unfamiliar, unexpected, or unplanned and either not visible or not immediately comprehensible for practitioners who carry out the safety-critical work. Linear interactions are those that happen in expected and familiar production or maintenance sequences and those that are quite visible and understandable, even if unplanned. Complex interactions, in contrast, produce unfamiliar sequences, or unplanned and unexpected sequences, and are either not visible or not immediately comprehensible. Failures in interactively complex systems (like a human body) can cascade in ways that confound the practitioners managing them, making it difficult for them to meaningfully intervene.

In addition to being either linearly or complexly interactive, systems can be loosely or tightly coupled. They are tightly coupled if they have more time-dependent processes (meaning they cannot wait or stand by until attended to), have sequences that are invariant (the order of the process cannot be changed), and have little slack (e.g., things cannot be done twice to get it right). Rail transport is a rather linear system but is tightly coupled. In contrast, an example of a system that is interactively complex but not tightly coupled is a university education. It is interactively complex because of specialization, limited understanding, number of control parameters, and so forth. But the coupling is not tight. Delays or temporary halts in education are possible, different courses can often be substituted for one another (as can a choice of instructors), and there are many ways to achieve the goal of obtaining a degree.

For normal accident theory, the two dimensions (interactive complexity and coupling) presented a dilemma. A system with high interactive complexity can only be effectively controlled by a decentralized organization. The reason is that highly interactive systems generate the sorts of nonroutine situations that resist standardization (e.g., through clinical guidelines, which is a form of centralized control fed forward into the operation). Instead, the organization has to allow operational practitioners considerable discretion and leeway to act as they see fit based on the situation. It also has to encourage direct interaction among operational practitioners, so that these can bring together the different kinds of expertise and perspectives necessary to understand and attack the problem.

A system with tight couplings, on the other hand, can in principle only be effectively controlled by a highly centralized organization because tight coupling demands quick and coordinated responses. Centralization in this sense does not mean hierarchical or bureaucratic but rather tightly coordinated and swiftly in place. Disturbances that cascade through a system cannot be stopped quickly if a team with the right mix of expertise and backgrounds needs to be assembled from the ground up. Tight coordination (e.g., through procedures, emergency drills, or even automatic shutdowns or other machine interventions) may be necessary to arrest such cascades quickly. Also, conflicts between different well-meaning interventions can make the situation worse, which means that activities oriented at arresting the failure propagation need to be extremely tightly coordinated.

The problem, in theory at least, is that an organization cannot be centralized and decentralized at the same time. So, the dilemma arises if a system is both interactively complex and tightly coupled, which medicine is on various occasions. A necessary conclusion for normal accidents theory is that systems that are both tightly coupled and interactively complex can therefore not be controlled effectively. In practice, the various attempts at implementing both guidelines and quick response teams (Berwick, Calkins, McCannon, & Hackbarth, 2006) showed that healthcare is trying to manage the contradiction. A mix of centralization (guidelines) and decentralization (quick response teams) can be applied to make propagating patient problems more manageable.

These examples show that humans are hardly the recipient victims of structural complexity and coupling alone. The definition of interactive complexity actually involves both human and system, to the point at which it becomes hard to see where one ends and the other begins. For example, interactions cannot be unfamiliar, unexpected, unplanned, or not immediately comprehensible in some system independent of the people who need to deal with them (and to whom they are either comprehensible or not). One hallmark of expertise, after all, is a reduction of the degrees of freedom that a decision presents to the problem solver and an increasingly refined ability to recognize patterns of interactions and knowing what to do primed by such situational appreciation. Interactive complexity thus cannot be a feature of a system by itself but always has to be understood in relation to the people (and their expertise) who have to manage that system. This also means that the categories of complexity and coupling are not as independent as normal accident theory suggests.

Another problem arises when complexity and coupling are treated as stable properties of a system because this misses the dynamic nature of work in healthcare and the ebb and flow of cognitive and coordinative activity to manage it. Coupling between units, components, or processes, for example, can go from loose to tight, depending on the circumstances.

Going Solid

Cook and Rasmussen (2005) showed how a hospital can gyrate from loose to tight coupling, depending on resource availability and patient loads. Clinical practitioners, they explained, are familiar with "bed crunch" situations during which a busy unit such as a surgical intensive care unit (ICU) becomes the operational bottleneck within a hospital. Other units in the hospital can usually buffer the consequences of a localized bed crunch by absorbing workload or deferring transfers.

But Cook and Rasmussen (2005) reported situations for which the entire hospital is saturated with work, creating a systemwide bed crunch. They called this "going solid." For example, in one case a surgical procedure was cancelled after induction of anesthesia because a scheduled transfer of another patient out of the ICU was made impossible by deterioration of that patient's condition. The anesthetic was started because it had become routine to begin surgery in anticipation of resources becoming available rather than waiting for them to be available. The patient would have required an ICU bed for recovery after the procedure, and the practitioners elected to halt the operation when it became apparent that no ICU bed would be available.

In a going-solid scenario at a much larger scale (Cook and Rasmussen, 2005, p. 131), a circulating nurse called the recovery room in anticipation of bringing a patient (patient 1) to it near the end of a routine scheduled surgical procedure. The recovery room placed the transfer from the operating room "on hold" because all the recovery room locations were filled by patients. Among these was patient 2, who should have been transferred from the operating room directly to an ICU bed. Patient 2 was in the recovery room because there was no ICU bed available: It was occupied by patient 3, whose condition would allow transfer to the regular ward but the regular ward bed was occupied by patient 4, who was ready for discharge but was awaiting arrival of a family member to transport him to his home.

Bed occupancy within the hospital had been at saturation for both ICU and regular ward beds for several weeks. The high-occupancy situation was managed by nurses and administrators by pairing new postoperative admissions with anticipated patient discharges, matching expected discharge and expected end of surgery times. Senior hospital management became involved in moment-to-moment decision making about bed allocation, surgical procedure starts, and intrahospital patient transfers. Managers also sought increased efficiency of resource use, mainly through direct inquiries about patient status. A new administrative nursing position was established to centralize and rationalize bed resources. The system remained solid for approximately 5 weeks (Cook & Rasmussen, 2005).

During periods of crisis, or high demand, a system can become more difficult to control as couplings tighten and interactive complexity momentarily deepens. It renders otherwise visible interactions less transparent, less linear, creating interdependencies that are harder to understand and more difficult to correct. This can become especially problematic when important routines are interrupted, coordinated action breaks down, and misunderstandings occur (Weick, 1990). The opposite also applies. Contractions in complexity and coupling can be met in centralized and decentralized ways by people responsible for the safe operation of the system, creating new kinds of coordinated action and newly invented routines. This theme is revisited in the section on practical drift.

Normal accident theory has a pessimistic view for those who believe in scientific-bureaucratic medicine. The ability of healthcare organizations to protect themselves against normal adverse events can fall victim to the very interactive complexity and tight coupling it must contain. Plans for emergencies, for example, are intended to help the organization deal with unexpected problems and developments. But, sometimes these plans are chiefly designed to be maximally persuasive to regulators, board members, surrounding communities, lawmakers, and opponents of the technology or its practitioners, and resulting guidelines can become wildly unrealistic. Clarke and Perrow (1996) called them "fantasy documents" that fail to cover most possible accidents, lack any historical record that may function as a reality check, and are quickly based on obsolete contact details, organizational designs, function descriptions, and divisions of responsibility. The problem with such fantasy documents is that they can function as an apparently legitimate placeholder that suggests that everything is under control. It inhibits the commitment of the organization to continually review and reassess its ability to deal with risk.

RISK AS THE GRADUAL ACCEPTANCE OF THE ABNORMAL

Man-made disaster theory, the Swiss cheese model, and normal accidents theory were all inspired by mishaps that called for additional or new theorizing. The mishaps presented researchers with new kinds of data that did not fit any existing framework. Accidents in 1986 as well as 1993 did this again, giving rise to new theorizing about adverse events. The theories that emerged shed particular light on processes of organizational adaptation and the flexibility of social interpretations of risk in the face of uncertainty. They added a richness to scientific knowledge about accidents, a new way of conceptualizing what actually went on in the interior of normal organizations before they produced an adverse event. They began to fill in the blanks of Barry Turner's (1978) incubation period.

THE NORMALIZATION OF DEVIANCE

In 1986, the space shuttle *Challenger* broke apart 73 seconds after the launch of its tenth mission, resulting in the death of all seven crew members, including a civilian teacher. The accident was investigated by a presidential commission and made to fit a structural account. Organizational deficiencies (holes in the layers of defense) had led to a tendency of engineers and managers not to report upward. These holes needed to be fixed by changes in personnel, organization, or indoctrination. The report placed responsibility for "communication failures" with middle managers, who had overseen key decisions and the implementation of inadequate rules and procedures (Vaughan, 2005). The presidential commission traced their managerial wrongdoing to a combination of enormous economic strains *and* operational expectations put on the space shuttle program.

Space shuttles had been designed and built (and sold to the taxpayer) as a bus-like recyclable and operational technology for trips into space. In reality, production expectations and operational capacities were far apart. To try to live up to launch schedules despite innumerable technical difficulties, middle management allowed rule violations and contributed to the silencing of those with bad news. Information about hazards (among others, an O-ring blow-by problem in solid rocket boosters that eventually led to the *Challenger* accident) was suppressed to stick to the launch schedule. On the eve of the launch of *Challenger*, managers consciously decided to take the risk of launching a troubled design in exceptionally low temperature. A teacher was going to be on board, and the president was going to hold his State of the Union address during her trip in space. The presidential commission exonerated top administrators, arguing that they did not have the information. It blamed middle management, however, because they did have all the information. They were warned against the launch by their engineers but decided to proceed anyway.

The account of the presidential commission was consistent with rational choice theory, in which humans are considered to be perfectly rational decision makers. Middle management at NASA (National Aeronautics and Space Administration) was assumed to have had exhaustive access to information for their decisions, as well as clearly defined preferences and goals about what they wanted to achieve. Despite this, they decided to go ahead with the launch anyway, to gamble. They were amoral calculators, in other words.

Richard Feynman, physicist and maverick member of the presidential commission, wrote a separate, dissenting appendix to the report on his own conclusion. One issue that disturbed him particularly was the extent of structural disconnects across NASA and contractor organizations. These uncoupled engineers from operational decision making and led to an inability to convince managers of (even the utility of) an understanding of fundamental physical and safety concepts. Feynman's conclusions very much fit the fears of those concerned with the conversion to scientific-bureaucratic medicine (see Chapter 1). In such a conversion, professional accountability and the dominance of a technical culture and expertise are gradually replaced with bureaucratic accountability, by which administrative control is centralized at the top, and the focus of decision makers is trained on business ideology and the meeting of political expectations.

As explained in Chapter 2, the premise of fully rational decision makers was replaced in the 1970s by the notion of local rationality. Middle managers would not have possessed everything on which to base rational decisions. People do not have full knowledge of all relevant information, possible outcomes, relevant goals. It would be impossible. What they do instead makes sense to them given their local goals, their current knowledge, and their focus of attention at the time. So, Vaughan (1996) argued, if there was amoral calculation and conscious choice to gamble on the part of middle management, then it was so only in hindsight.

It was not difficult to agree on the basic facts. NASA had been under extraordinary production pressure and economic constraints in the years leading up to the accident. Its reusable spacecraft had been sold on the promise of cost-effectiveness and won an endorsement from the Air Force (which put outlandish payload expectations on the design in return). To break even, it had to make 30 flights per year, and even this was considered a conservative estimate. Funds eventually allocated to NASA were half of what had been requested. NASA never got all the funds it needed, and the space shuttle was moved from a developmental and testing flight regime into an operational one early. The design was one big trade-off. Reductions in development costs that were needed to secure funding were attained by exporting cost into higher future operating costs. True launch costs ballooned up to 20 times the original estimates, even as competition was growing from commercial spaceflight in other countries, and a third of the NASA workforce had disappeared as compared to the *Apollo* years. Only nine missions were flown in 1985, the year before *Challenger*, and even the launch of *Challenger* was delayed multiple times.

It was easy to reach agreement on the notion that production pressure and economic constraints are a powerful combination that could lead to adverse events. But how? Was it a matter of conscious trade-offs, of actively sacrificing safety and making immoral calculations that gambled with people's lives? Or, was something else going on? Researchers started to dig into the social processes that slowly but surely converted the bad news of technical difficulties and operational risk into something that was normal, acceptable, expected, manageable (Starbuck & Milliken, 1988; Vaughan, 1999, 2005).

These are important questions at the heart of organizational risk management, operational work, and technological development—and in healthcare. The development and fielding of linear particle accelerators, for example, or the embrace of bar coding in blood testing and medication delivery, computerized medical records, or the implementation of a computerized operating room scheduling system are all examples of technologies that can help sustain the image of rational, managerial control over the cost of healthcare. If managers know that these technologies have problems, are they irrational and amoral in implementing them anyway? And are expert practitioners powerless and disenfranchised in the face of their decisions, like the engineers at NASA in the 1980s?

To the presidential commission, there was a linear relationship between scarcity, competition, production pressure, and managerial wrongdoing. Pressures developed because of the need to meet customer commitments, which translated into a requirement to launch a certain number of flights per year and to launch them on time. Such considerations may occasionally have obscured engineering problems. To Vaughan (1996), something else was going on. She traced the risk assessments of those involved with the key technology in *Challenger* (the solid rocket booster) and the development of meanings by insiders as problems with the design unfolded and became evident. She decided that this history portrays "an incremental descent into poor judgment."

For Vaughan, seeing holes and deficiencies in hindsight was not an explanation of the generation or continued existence of those deficiencies. It would not help predict or prevent failure. Instead, the processes by which such decisions come about, and by which decision makers create their local rationality, are one key to understanding how safety can erode on the inside of a complex, sociotechnical system. Why did these things make sense to organizational decision makers at the time? Why was it all normal, why was it not worthy of reporting?

To Vaughan, production pressures played a huge role, but not in the way envisioned by the presidential commission. Rather, production pressures and resource limitations gradually became institutionalized, taken for granted. They became a normal aspect of the worldview every participant brought to organizational and individual operational decisions.

This characterization of resource constraints and production pressures is echoed in the actions of the urologist in Chapter 1 who did not have access to the necessary equipment, instead modifying an existing but nonstandard device in an attempt to continue with the operation. Lacking the right equipment was probably not something that was worth reporting as it was relatively normal. The need for improvisation was normal also, which is what can be expected of good doctors. What the physician was dealing with was normal, natural trouble. And, achieving operational success with the improvisation may have given everybody the impression that no extra risk was taken, or that risk was under control, and that in any case this could be tried again with an expectation of a similar good outcome. Operational success may ultimately give management the impression that the necessary equipment is not even "necessary" for the operation. Perfectly acceptable clinical results are achieved without it as well.

To Vaughan (1999), this social-technical dynamic, in which the idea of risk is continuously constructed and renegotiated at the intersection of the technical and the social, was at the heart of the incubation period for *Challenger*. Indeed, this dynamic is fundamental to the construction of risk in any safety-critical endeavor. She called it the normalization of deviance. This is how a work group's construction of risk can persist even in the face of continued and worsening signals of potential danger (Vaughan, 1996). The use of a morcellator in spleenectomy could be an example of this. Risk is believed to be under control because of the presence of basic competence. Signals of potential danger in one operation (e.g., a near miss of an artery) are acknowledged and then rationalized and normalized, leading to continued use under apparently similar circumstances. This repeats itself until something goes wrong, revealing the gap between how risk was believed to be under control and its actual presence in the operation:

- *Beginning the construction of risk: a redundant system.* The starting point for the safety-critical activity is the belief that safety is ensured, and risk is under control. Redundancies, the presence of extraordinary competence, or the use of proven technology can all add to the impression that nothing will go wrong. There is a senior surgeon who watches the operation through a laparoscope from a distance.
- *Signals of potential danger.* Actual use or operation shows a deviation from what is expected. An artery is almost grazed, for example. This creates some uncertainty and can indicate a threat to safety, thus challenging the original construction of risk.
- *Official act acknowledging escalated risk.* This may consist of a more senior physician taking over the operation, only to hand back the technology to the more junior doctor after stabilizing the situation.
- *Review of the evidence.* After the operation, discussions may ensue about who did what and how well things went. This is not necessarily the case after all operations; in fact, it could be the kind of standardized debrief that is hardly institutionalized in medicine yet (see Chapter 6).
- *Official act indicating the normalization of deviance, accepting risk.* The escalated risk can be rationalized or normalized as a complication, as an anomaly, or rationalized and localized as a deviant anatomy that was unique to this patient, for example. If there is a more formal review, it might show that the operating doctor was formally certified to use the technology (independent of actual experience with it) because of the basic medical qualification. Redundancy was ensured through multiple levels of senior presence during the operation. And, the technology itself had undergone various stages of testing and revision before being fielded. All

these factors contribute to a conclusion that any risk was duly assessed and under control.
* *Continued operation.* The technology will be used again in the next operation because nothing went wrong, and a review of the risks has revealed that everything is under control.

The key to the normalization of deviance is that this process, this algorithm, repeats itself, and that successful outcomes keep giving the impression that risk is under control. Success typically leads to subsequent decisions, setting in motion a steady progression of incremental steps toward greater risk. Each step away from the previous norm that meets with empirical success (and no obvious sacrifice of safety) is used as the next basis from which to depart just that little bit more again. It is this incrementalism that makes distinguishing the abnormal from the normal so difficult. If the difference between what "should be done" (or what was done successfully yesterday) and what is done successfully today is minute, then this slight departure from an earlier established norm is not worth remarking or reporting. Incrementalism is about continued normalization: It allows normalization and rationalizes it.

> Experience generates information that enables people to fine-tune their work: fine-tuning compensates for discovered problems and dangers, removes redundancy, eliminates unnecessary expense, and expands capacities. Experience often enables people to operate a socio-technical system for much lower cost or to obtain much greater output than the initial design assumed. (Starbuck & Milliken, 1988, p. 333)

Normalizing deviance is about fine-tuning, adaptation, and increments. Decisions that are seen as "bad decisions" after an adverse event (like using unfamiliar or nonstandard equipment) seemed like perfectly good or reasonable proposals at the time. No amoral, calculating people were necessary to explain why things eventually went wrong. All it took was normal people in a normal organization under the normal sorts of pressures of resource constraints and production expectations. These are people with normal jobs and real constraints but no ill intentions, whose constructions of meaning coevolved relative to a set of environmental conditions, and they tried to maintain their understanding of those conditions. Because of this, said Vaughan (1996),

> Mistake, mishap, and disaster are socially organized and systematically produced by social structures. No extraordinary actions by individuals explain what happened: no intentional managerial wrongdoing, no rule violations, no conspiracy. The cause of the disaster was a mistake embedded in the banality of organizational life and facilitated by an environment of scarcity and competition, an unprecedented, uncertain technology, incrementalism, patterns of information, routinization, organizational and interorganizational structures. (p. xiv)

Vaughan's (1996) analysis showed how production pressures and resource constraints, originating in the environment, become institutionalized in the practice of people. These pressures and constraints have a nuanced and unacknowledged yet pervasive effect on organizations and the decision making that goes on in them. This is why harmful outcomes can occur in activities and organizations constructed precisely to prevent them (indeed, like healthcare). The structure of production pressure and resource scarcity becomes transformed into organizational mandates and affects what individual people see as normal, as rational, as making sense at the time.

There is a relentless inevitability of mistake, the result of the emergence of a can-do culture. This is created as people interact in normal work settings where they normalize signals of potential danger so that their actions become aligned with organizational goals. This normalization ensures that they can tell themselves and others that no undue risk was taken, and that in fact the organization benefited: The operation was completed, no resources were used unnecessarily or wasted. Through repeated success, this work group culture persists; it rationalizes past decisions while shaping future ones. But in the end, the incrementalism of those decisions contributes to extraordinary events.

How to Prevent the Normalization of Deviance

Vaughan (1996) was not optimistic. Adverse events, she was forced to conclude, are "produced by complicated combinations of factors that may not congeal in exactly the same way again" (p. 420). Conventional explanations that focus on managerial wrongdoing leave a clear recipe for intervention (indeed the one supplied by the presidential commission). Get rid of the really bad managers, change personnel around, amend procedures and rules. But it is unclear whether this might prevent a repetition of processes of normalization:

> We should be extremely sensitive to the limitations of known remedies. While good management and organizational design may reduce accidents in certain systems, they can never prevent them. ... System failures may be more difficult to avoid than even the most pessimistic among us would have believed. The effect of unacknowledged and invisible social forces on information, interpretation, knowledge, and—ultimately—action, are very difficult to identify and to control. (p. 416)

Indeed, the space shuttle *Columbia* accident in 2003 showed how NASA history was able to repeat itself: Foam strikes to the wing of the space shuttle had become normalized as acceptable flight risk and converted into a maintenance problem rather than a flight safety problem (Columbia Accident Investigation Board, 2003).

Vaughan's (1996) analysis revealed a more complex picture behind the creation of such accidents and the difficulty in learning from them. It shifts attention from individual causal explanations to factors that are difficult to identify and untangle, yet have a great impact on decision making in organizations (p. xv). Vaughan's is a more frightening story than the historically accepted interpretation. Its invisible processes and unacknowledged influences tend to remain undiagnosed and therefore elude remedy.

The solution to risk, if any, is to ensure that the organization continually reflects critically on and challenges its own definition of normal operations and finds ways to prioritize chronic safety concerns over acute production pressures. But how is this possible when the definition of *bad news* is something that gets constantly renegotiated as success with improvised procedures or imperfect technology accumulates? Such past success is taken as a guarantee of future safety. Each operational success achieved at incremental distances from the formal, original rules or procedures or design requirements can establish a new norm. From there, a subsequent departure is again only a small incremental step.

From the outside, such fine-tuning constitutes incremental experimentation in uncontrolled settings. On the inside, incremental nonconformity is an adaptive response to scarce resources and production goals. This means that departures from the routine become routine. Seen from the inside of people's own work, deviations become compliant behavior. They are compliant with the emerging, local ways to accommodate multiple goals important to the organization (maximizing capacity utilization but doing so safely; meeting not only technical or clinical requirements but also deadlines).

In making these trade-offs, however, there is a feedback imbalance. Information on whether a decision is cost-effective or efficient can be relatively easy to obtain. One good clinical outcome is measurable and has immediate, tangible benefits. How much is or was borrowed from safety to achieve that goal, however, is much more difficult to quantify and compare. Each consecutive empirical success (the operation is successful, the theater is now available again for the next procedure) seems to confirm that fine-tuning is working well: The system can operate safely yet more efficiently.

As Weick (1993) pointed out, however, safety in those cases may not at all be the result of the decisions that were or were not made but rather an underlying stochastic variation that hinges on a host of other factors, many not easily within the control of those who engage in the fine-tuning process. Empirical success, in other words, is no proof of safety. Past success does not guarantee future safety. Borrowing increasingly from safety may go well for a while, but you never know when you are going to hit a problem. This moved Langewiesche (1998) to say that Murphy's law is wrong: Everything that can go wrong usually goes right, and then we draw the wrong conclusion.

The uneasy tension between bureaucratic accountability and managerial infatuation with standardization and control on the one hand and professional, technical culture on the other (see Chapter 1) may present another possibility for intervention. One solution would be to empower people closer to the technology to speak up, to collectively consult and cast deciding votes (Pronovost & Vohr, 2010). There are a number of advantages associated with this. Taking expertise seriously and sensitizing organizational decision making to voices of dissent and minority opinion are prescriptions of high-reliability theory (see elsewhere in this chapter). But, as normal accident theory pointed out, such empowerment can exacerbate the conflict between centralization and decentralization. Procurement decisions and the implementation of new technology in hospitals increasingly require centralization because of the typical size of such projects, their organization-wide ramifications, and budgetary implications. Decentralization can undermine the success and justification entirely, as with the computerized operating room scheduling system in Chapter 1.

STRUCTURAL SECRECY AND PRACTICAL DRIFT

Large organizations like hospitals are made up of an enormous number of levels, departments, disciplines, and specializations. In the wake of an adverse event (particularly one that generates media attention and becomes a celebrated case), it is easy to assert that some people may have intentionally concealed information about the difficulties with a particular person or technology, thus preventing the administration from intervening and addressing the problem. But that is often only obvious in hindsight. Such secrecy is often simply a structural by-product of how work is organized, particularly inside large, bureaucratic organizations (like a hospital).

March, Cohen, and Olsen (1988) developed an important perspective on the functioning of people inside such organizations. They saw organizations as natural, open systems, not rational, closed systems in which people can accomplish work in a way that is immune from everything that makes us human. They are "natural" like all social groups. People actively pursue goals of narrow self-interest and may prioritize things that only benefit their local group. This can well be to the detriment of official goals of openness, profit, or production. The idea that organizations are capable of inculcating a safety orientation among its members through recruitment, socializing, and indoctrination is met with great skepticism. Then, the organizations in which this happens are not closed off to the environment. They are "open" in the sense that they are constantly perfused by forces from outside: political, economic, social. These forces enter into all kinds of decisions and preferences that are expressed by various groups and people on the inside.

March and colleagues (1988) argued that organizations cannot make up for the limited, local rationalities of the people inside them. Instead, organizations themselves have a limited or local rationality because they are made up of people put together. In fact, organizations function more like "organized anarchies" or "garbage cans." Problems are ill defined and often unrecognized; decision makers have limited information, shifting allegiances, and uncertain intentions. Solutions may be lying around, actively searching for a problem to attach themselves to. Possible moments for rational choice are opportunistic and badly coordinated. People often hardly prepare well for these moments (e.g., budget or board meetings), yet the organization is expected to produce behavior that can be called a decision anyway.

This provocative and pessimistic vision of organizational life stands in contrast with high-reliability theory (see discussion in a separate section of this chapter), which argues that organizations in fact can make up for the local rationalities of the people inside them. It just takes better, well, organization. But, for March and colleagues (1988), these hopes were in vain. For them, organizations like hospitals operate on the basis of a variety of inconsistent and ill-defined preferences. This leads to goal conflicts and diversions of aims and direction in various departments. The organization may not even know its preference until after a decision has been made. Also, organizations use unclear technologies in their operations, which means that people inside them might not even really understand why some things work and some things do not (see the section on "unruly" technology). Finally, organizational life is characterized by an extremely fluid participation in decision-making processes. People come and go, with various extents of commitment to organizational vision and goals.

Some pay attention while they are there; others do not. Key meetings may be dominated by biased, uninformed, or even uninterested personnel.

With this vision, it is not surprising that an organization is hardly capable of arriving at a clear understanding of its risks and commitments for how they should be managed. Structural secrecy, with participants not knowing about what goes on in other parts of the organization, is a normal by-product of the bureaucratic organization and social nature of complex work. Jensen (1996) described it as such:

> We should not expect the experts to intervene, nor should we believe that they always know what they are doing. Often they have no idea, having been blinded to the situation in which they are involved. These days, it is not unusual for engineers and scientists working within systems to be so specialized that they have long given up trying to understand the system as a whole, with all its technical, political, financial, and social aspects. (p. 368)

By structural secrecy, Vaughan (1996) meant the way that patterns of information, organizational structure, processes, and transactions all undermined people's attempts to understand what was going on inside the organization. As organizations grow larger, most actions are no longer directly observable. Labor is divided between subunits and hierarchies and geographically dispersed. Bad news can remain localized, seen as normal where it is, and not be subjected to any credible outside scrutiny. Distance, both physical and social, interferes with efforts from those at the top to know what is going on and where:

> Specialized knowledge further inhibits knowing. People in one department or division lack the expertise to understand the work in another or, for that matter, the work of other specialists in their own unit. ... Changing technology also interferes with knowing, for assessing information requires keeping pace with these changes—a difficult prospect when it takes time away from one's primary job responsibilities. To circumvent these obstacles, organizations take steps to increase the flow of information—and hypothetically, knowledge. They make rules designating when, what, and to whom information is to be conveyed. Information exchange grows more formal, complex, and impersonal, perhaps overwhelmingly so, as organizations institute computer transaction systems to record, monitor, process and transmit information from one part of the organization to another. Ironically, efforts to communicate more can result in knowing less. (p. 250)

With clinical work divided along lines of specialty or departmental responsibility, discontinuities are a normal, expected by-product. And with partitions between administration or management and clinical practitioners, more fissures and gaps open. It is precisely in these gaps, these crevasses, that problems can arise, that slippage in intention and communication easily occur (R. I. Cook, Render, & Woods, 2000). The secrecy is built in. Indeed, as predicted by Vaughan (1996), computerized systems intended to increase information flow (e.g., computerized patient records) import a whole set of new problems while not necessarily resolving the fundamental discontinuities between rank, hierarchy, specialization, and interest.

Recall the local rationality principle from Chapter 2. People do what locally makes sense to them given their goals, knowledge, and focus of attention in that setting. Scott Snook (2000), who produced a revisionist account of the mistaken shoot down of two U.S. Black Hawk helicopters by U.S. fighter jets over northern Iraq in 1993, called it "practical action" (Vaughan 1996, p. 182). This is "behavior that is locally efficient, acquired through practice, anchored in the logic of the task, and legitimized through unremarkable repetition."

Using a shortcut in the bar coding of medications or blood could be an example. For instance, in one ward, nurses had affixed particular bar codes behind each bed so that they could easily scan a code while at the bedside without actually having to go to the medication cabinet outside the room. Successfully adapting a piece of technology not intended for that surgical procedure may be another example. As with Vaughan's normalization of deviance, operational success with such adapted procedures is one of the strongest motivators for doing it repeatedly. Recall that this is also sustained because of feedback asymmetry: There are immediate and acute productive gains and little or no feedback about any gathering danger, particularly if the procedure was successful.

Plans and procedures, then, for dealing with potentially safety-critical tasks are always subject to local modification. Those charged with implementing and following them find ways to work around aspects of plans or protocols that impede fluid accomplishment of their tasks. What can happen then is that these local adaptations drift further from the original rationale for implementing the tighter procedure or protocol in the first place. This can occur in large part because of structural secrecy. Little may be seen or known, either vertically or horizontally in an organization, about how local groups adapt practice. Drift occurs in the seams between departments, professions, specializations, or hierarchies. It slowly but steadily decouples localized realizations of the activity from the original centralized task design. The local deviance from centralized impositions or expectations is slow and sensible enough (and successful enough) to be virtually unnoticeable, particularly from the inside of that local work group. It becomes acceptable and expected behavior; the deviant becomes the norm. Risk is not seen because it is packed away in the gradual acceptance of deviance (and its conversion into normality) within local work groups.

In loosely coupled situations (remember normal accidents theory), this may not be a problem. Routine situations, in which there is time to reflect, to stop, to start over, can easily absorb adaptations and improvisations, even if these are not really coordinated across professional or hierarchical boundaries. People or groups may not have to shed their locally evolved rational actions and habits because it does not really have negative consequences. There is always margin, or slack, to recover and repair misunderstanding or confusion.

But, in a rapidly deteriorating situation, coupling becomes tighter. Time can be of the essence; things have to be done in a fixed, rapid order (if the bag valve mask does not work, you cannot ventilate without intubating, and if that does not work, you cannot do a tracheotomy without anesthetic), and substitution of material, expertise, or protocols is not possible (you might *really* need an anesthetist to solve this particular problem). If various local practices have drifted apart over time, by not coordinating their adaptations or improvisations with others, it may become really difficult to act smoothly and successfully in a tightly coupled situation. When different practitioners or teams come together to solve a problem that is outside the routine, and for which they usually do not see each other, the effects of drift quickly come to the surface. Some people's roles may have subtly shifted ("But *she* used to do this, why are you looking at me?"), material may have been sorted and placed elsewhere ("Where did we put that thing?"), objects may have begun to be called by different names ("You want *what*?"). Risk becomes evident only when different locally adapted practices meet.

As Perrow (1984) pointed out, a situation with tight couplings can only be effectively controlled by highly centralized organization because tight coupling demands quick and coordinated responses. People need to know what they can expect from others, even if they do not work together often or only work together in emergency situations. Practical drift is fine for situations with loose coupling. The effects, however, can be disastrous in situations that are tightly coupled. A rapidly deteriorating situation cannot be stopped quickly if a team with the right mix of expertise and backgrounds needs first to figure out how to work together. Centralization (e.g., through procedures, emergency drills, standardized roles and phraseology) is necessary to arrest such cascades quickly. It is not surprising that Berwick argued that the deployment of rapid response teams at the first sign of patient decline would be a good investment in safety (Berwick et al., 2006). Rapid response teams that have some form of centralized control and close coordination are better capable of dealing with tightly coupled situations that may follow from the first signs of patient decline.

CLINICAL GUIDELINES AND PATIENT SAFETY

Some might see Snook's (2000) results on the dangers of practical drift as a strong reason to forge ahead with implementing various forms of scientific-bureaucratic medicine (see Chapter 1), arguing that medical work should not be beyond the reach of written rules and externally imposed scripts. One risk is that process implementation or improvement becomes an end in itself, and that any evidence that links it to greater patient safety or organizational efficiency is purely imagined. The assumption is that order and stability in healthcare are achieved rationally and mechanistically and that control should be implemented vertically (e.g., through task analyses that produce prescriptions of work to be carried out). This may indeed be the belief (Dekker, 2003):

- Guidelines represent the best thought-out and thus the safest way to carry out a job. They are based on evidence.
- Guideline following is mostly simple if-then rule-based mental activity: If this situation occurs, then this algorithm (e.g., clinical protocol) applies.
- Safety results from practitioners following guidelines. Risk comes from practitioners not following the guidelines.
- For progress on safety, organizations must invest in practitioners' knowledge of guidelines and ensure that they are followed.

In the wake of an adverse event, it can be tempting to introduce new guidelines, change existing ones, or enforce stricter compliance. This does not just happen in healthcare. The Air Force also does this. For example, shortly after a fatal shoot down of the two U.S. Black Hawk helicopters over northern Iraq (Snook, 2000), the following occurred:

> Higher headquarters in Europe dispatched a sweeping set of rules in documents several inches thick to "absolutely guarantee" that whatever caused this tragedy would never happen again. (p. 201)

It is a common, but not typically satisfactory, reaction. Introducing more guidelines does not necessarily avoid the next adverse event, and exhortations to follow rules more carefully do not necessarily increase compliance or enhance safety (or managerial credibility, for that matter). To be sure, guidelines with the aim of standardization can play an important role in shaping safe practice (see Chapter 6). Commercial aviation is often held up as prime example of the powerful effect of standardization on safety. But even there, ambiguity abounds, and evidence exists that procedures are a more problematic category of human work.

Local adaptation of all forms of written or imposed guidance is a necessary and inevitable feature of any complex social system. Departures from some guideline or routine may at any one moment seem to occur because people are not motivated to do otherwise or because people have become reckless and focused only on production goals. But, departures from the routine can become the routine as a result of a much more complex picture that blends organizational or team preferences (and how they are communicated), earlier success, peer pressure, and compliance with implicit expectations.

Guidelines can typically not present any close relationship to situated action because of the unlimited uncertainty and ambiguity involved in the activity. Guidelines, because of their very generalization, are often seen as alien to the practice of medicine. If guidelines were a perfect match for the actual task, there would not be the need for a human to carry them out; computers, automation, and robotics could do so more reliably. Medicine fits what Brian Wynne (1988) called "unruly technology." *Unruly* means that something is disorderly and not fully amenable to discipline or control (like human anatomy or physiology).

Unruly is a better word than uncertain. *Uncertain* means unknown. The whole point of learning about medicine is to reduce those unknowns. We run the cases, do the differentials, perhaps even build the computer simulations, so that we know, for example, under which circumstances something will stop functioning. But, independent of the reduction of uncertainty, the systems that medicine deals with will remain unruly—leaving practitioners to deal with the messy, not-so-governable interior of complex systems and situations that are diverse and resistant to standardized responses. Unruly technology is characterized by ambiguity, a lack of structure, and deviation from standard (anatomical) specifications and (physiological) operating standards. From this perspective, a sick patient *always* looks like an adverse event waiting to happen. Yet, as Paget (2004) reminded us, this is "normal."

A crucial skill, then, involves finding a practical balance between the universality of assumptions and their contextualization. One way in which this occurs is that rules emerge from practice and experience rather than precede them. Guidelines, in other words, end up following work instead of specifying action beforehand. Medical practice, even in the total absence of guidelines, is rule abiding, but those rules are practical rules, unwritten ones, heuristics, experience driven. They can be operating standards that consist of numerous ad hoc judgments and assumptions grounded in evolving practice (Hugh & Dekker, 2008). Wynne (1988) captured the essence of such unruly technology:

> Beneath a public image of rule-following behavior and the associated belief that accidents are due to deviation from those clear rules, experts are operating with far greater levels of ambiguity, needing to make uncertain judgments in less than clearly structured situations. The key point is that their judgments are not normally of the kind "how do we design, operate and maintain the system according to *the* rules?" Practices do not follow rules. Rather, rules follow evolving practices. (p. 153)

Unruly technology can be brought under control in various ways (but never perfectly so). One way is through numbers (e.g., Apgar score for newborns, standard for oxygen saturation). But, numbers are just a starting point. From that point, medical work is guided by a system of flexible (and mostly unwritten) rules that are tailored and retailored to suit an evolving knowledge base. The patients themselves may turn into their own models as knowledge and guidelines are developed to deal with their unruly technology. The observation of patients' functioning before or after an intervention, and especially their malfunctioning, on a real scale is required as a basis for further clinical judgment and knowledge (Weingart, 1991).

Safety, then, is not the result of rote rule following; it is the result of people's insight into the features of situations that demand certain actions and of people being skillful at finding and using a variety of resources (including written guidance) to accomplish their goals. This suggests a second model for guidelines and patient safety:

- Guidelines are resources for action. Guidelines themselves are not the work. Guidelines do not specify all circumstances in which they apply. Guidelines cannot dictate their own application. Guidelines can, in themselves, not guarantee safety.
- Applying guidelines successfully across situations is a substantive and skillful cognitive activity. It takes experience and expertise.
- Safety results from practitioners being skillful at judging when and how (and when not) to adapt guidelines to local circumstances.
- For progress on safety, organizations must monitor and understand the reasons behind the gap between guidelines and practice. In addition, organizations must develop ways that support practitioners' skills at judging when and how to adapt.

Instituting managerial policies that penalize or stigmatize the departure from guidelines (by calling such noncompliance a "violation") are not something on which medical practitioners and managers can likely ever reach agreement. Calling the gap between guideline and practice a violation is simply a choice. It is a choice that has moral overtones and can be tremendously underinformed about the messy details of what it means to practice with unruly technology.

Rather, the discovery of a gap between guidelines and practice should lead to the recognition that it is often compliance (not deviance or violations) that explains practitioners' behavior. They comply with norms and local expectations that have evolved over time. They comply with unwritten rules and operating standards that probably make good local (and clinical) sense. What practitioners are doing is likely entirely reasonable in the eyes of those on the inside of the situation, given the pressures and priorities operating on them and others doing the same work every day.

The management of a gap between guidelines and practice should involve finding out what organizational history or pressures exist behind these routine departures from written guidance and which other goals help shape the new norms for what is acceptable risk and behavior. It involves understanding that the rewards of departures from the routine are probably immediate and tangible, and that the potential risks (how much was borrowed from safety to achieve those goals?) are unclear, unquantifiable, or even unknown. It involves realizing that continued absence of adverse consequences may confirm people in their beliefs (in their eyes, justified) that their behavior was safe while also achieving other important system goals.

RISK AS A MANAGERIAL OR CONTROL PROBLEM

The previous three schools of thought have conceptualized risk as energy to be contained with more or better barriers, as a structural property of complex systems, and as the gradual acceptance of the abnormal. Their prescriptions of how to deal with risk have ranged from putting in more or better barriers, to reducing the complexity of the system, to getting better at understanding and influencing what is normal or standard practice. The final conceptualization of risk here is as a managerial or control problem.

It is a potentially constructive view that in fact takes all previous conceptualizations seriously. Yes, risk can lie in dangerous energy that needs containing, it can be a structural property of the complex system, and it can be associated with normative and practical drift. But that does not mean that people inside an organization are at the mercy of risk. There is much that they can do about it. The solution to risk lies in a repertoire of organizational activities and managerial and professional commitments, countermeasures, and control checks. The first model that needs to be discussed here, then, is that of risk as a control problem. If risk is a control issue, then that opens up possible avenues for risk management. These then are discussed in light of what is known as high-reliability theory.

CONTROL THEORY

Control theory looks at adverse events as emerging from interactions among system components. It usually does not identify single causal factors but rather looks at what may have gone wrong with the operation of the system or organization of the hazardous technology that allowed an accident to take place. Safety, or risk management, is viewed as a control problem (Rasmussen, 1997), and adverse events happen when component failures, external disruptions, or interactions between layers and components are not adequately handled or when safety constraints that should have applied to the design and operation of the technology have loosened, or become badly monitored, managed, or controlled. Control theory tries to capture these imperfect processes, which involve people, societal and organizational structures, engineering activities, and physical parts. It sees the complex interactions between those—as did man-made disaster theory—as eventually resulting in an accident (N. Leveson, 2002).

Control theory sees the operation of hazardous processes (such as medication delivery or blood bank work) as a matter of keeping many interrelated components in a state of dynamic equilibrium. This means that control inputs, even if small, are continually necessary for the system to stay safe: Like a bicycle, it cannot be left on its own, or it would lose balance and collapse. A dynamically stable system is kept in equilibrium through the use of feedback loops of information and control. Adverse events are not seen as the result of an initiating event or root cause that triggers a linear series of events. Instead, adverse events result from interactions among components that violate the safety constraints on system design and operation. Feedback and control inputs can grow increasingly at odds with the real problem or processes to be controlled. Concern with those control processes (how they evolve, adapt, and erode) lies at the heart of control theory as applied to organizational safety.

Control theory says that the potential for failure builds because deviations from the original design assumptions of the system become increasingly rationalized and accepted. Adaptations occur, adjustments are made, and constraints are loosened in response to local concerns with limited time horizons. They are all based on uncertain, incomplete knowledge. Often, it is not even clear to insiders that constraints have become less tight as a result of their decisions in the first place, or that it matters if it is. And, even when it is clear, the consequences may be hard to foresee and judged to be a small potential loss in relation to the immediate gains. As Nancy Leveson (2002) put it, experts do their best to meet local conditions, and in the busy daily flow and complexity of activities, they may be unaware of any potentially dangerous side effects of those decisions. Note how these ideas echo concerns about local practical drift and structural secrecy that can engender and sustain erroneous expectations of users or system components about the behavior of others in the system.

A changed or degraded control structure eventually leads to adverse events. In control theoretic terms, degradation of the safety control structure over time can be due to asynchronous evolution, by which one part of a system changes without the related necessary changes in other parts. Changes to subsystems may have been carefully planned and executed in isolation, but consideration of their effects on other parts of the system, including the role they play in overall safety control, may remain neglected or inadequate. Asynchronous evolution can also occur when one part of a properly designed system deteriorates independent of other parts.

The more complex a system is (and, by extension, the more complex its control structure), the more difficult it can become to map out the reverberations of changes (even carefully considered ones) throughout the rest of the system. Control theory embraces a more complex idea of causation than the energy-to-be-contained models discussed (see also Chapter 8). Small changes somewhere in the system, or small variations in the initial state of a process, can lead to large consequences elsewhere.

Control theory helps in the design control and safety systems (particularly software based) for hazardous industrial or other processes (N. G. Leveson & Turner, 1993). When applied to organizational safety, control theory is concerned with how an erosion of a control structure allows a migration of organizational activities toward the boundary of acceptable safety performance.

Water Contamination Incident

Leveson and her colleagues (2003) applied control theory to the analysis of a water contamination incident that occurred in May 2000 in the town of Walkerton, Ontario, Canada. The contaminants *Escherichia coli* and *Campylobacter* entered the water system through a well of the Walkerton municipality, which operated the system through its Walkerton Public Utilities Commission (WPUC). Their control theoretic approach showed how the incident flowed from a steady (and rationalized, normalized) erosion of the control structure that had been put in place to guarantee water quality.

The proximate events were as follows: In May 2000, the water system was supplied by three groundwater sources: wells 5, 6, and 7. The water pumped from each well was treated with chlorine before entering the distribution system. The source of the contamination was manure that had been spread on a farm near well 5. Unusually heavy rains from May 8 to May 12 carried the bacteria to the well. Between May 13 and 15, a WPUC employee checked well 5 but did not take measurements of chlorine residuals, although daily checks were supposed to be made. Well 5 was turned off on May 15, and well 7 was turned on at that point. A new chlorinator, however, had not been installed on well 7, and the well was therefore pumping unchlorinated water directly into the distribution system. The WPUC employee did not turn off the well but instead allowed it to operate without chlorination until noon on Friday, May 19, when the new chlorinator was installed.

On May 15, samples from the Walkerton water distribution system were sent to a laboratory for testing according to the normal procedure. Two days later, the laboratory advised WPUC that samples from May 15 tested positive for *E. coli* and other bacteria. On May 18, the first symptoms of illness appeared in the community. Public inquiries about the water prompted assurances by WPUC that the water was safe. The next day, the outbreak had grown, and a physician contacted the local health unit with a suspicion that she was seeing patients with symptoms of *E. coli* infection.

In response to the lab results, WPUC started to flush and superchlorinate the system to try to destroy any contaminants in the water. The chlorine residuals began to recover. WPUC did not disclose the lab results. They continued to flush and superchlorinate the water through the following weekend, successfully increasing the chlorine residuals. Ironically, it was not the operation of well 7 without a chlorinator that caused the contamination; the contamination instead entered the system through well 5 from May 12 until it had been shut down on May 15.

Without waiting for more samples, a boil water advisory was issued on May 21. About half of Walkerton's residents became aware of the advisory on May 21, with some members of the public still drinking the Walkerton town water as late as May 23. Seven people died, and more than 2,300 became ill.

Although the proximate events could also be modeled using a sequence-of-events approach, pointing to the various errors, violations, and shortcomings in layers of defense, Leveson and colleagues decided to model the Ontario water quality safety control structure and show how it eroded over time to allow the contamination to take place. The safety control structure was intended to prevent exposure of the public to contaminated water, first by removing contaminants, second by public health measures that would prevent consumption of contaminated water.

In the case of Ontario, decisions had been taken to remove various water safety controls, or to reduce their enforcement, without an assessment of the risks. One of the important features that disappeared was feedback loops. As the other controls weakened or disappeared over time, the entire sociotechnical system moved to a state in which a small change in the operation of the system or in the environment (in this case, unusually heavy rain) could lead to a tragedy.

Well 5 had been vulnerable to contamination from the beginning. It was shallow, in an area open to farm runoff, and perched on top of bedrock with only a thin layer of topsoil around it. No extra approval for the well had been necessary, however, and it was connected to the municipal system as a matter of routine. No program or policy was in place to review existing wells to determine whether they met requirements or needed continuous monitoring.

Objections to the taste of chlorine in drinking water, WPUC employees who could safely consume untreated water from the wells, a lack of certification for water system operators, inexperience with water quality processes, and a focus on financial strains on WPUC led to the erosions of the control structure. A lack of government land use and watershed policy exposed this increasingly brittle structure to water heavily contaminated by hog and cattle farming. Budget and staff reductions by a new conservative government took a toll on environmental programs and agencies. A Water Sewage Services Improvement Act was passed in 1996, which shut down the government-run testing laboratories, delegated control of provincially owned water and sewage plants to municipalities, eliminated funding for municipal water utilities, and ended the provincial drinking water surveillance program. Farm operators were from that point to be treated with understanding if they were found in violation of livestock and wastewater regulations. No criteria were established to ensure the quality of testing or the qualifications or experience of private lab personnel, and no provisions were made for licensing, inspection, or auditing of private labs by the government. The resulting control structure was a hollowed-out version of its former self. It had become brittle and vulnerable to an unusual perturbation (like massive rainfall), lacking the resilience or redundancies to stop the problem or recover quickly from it.

For control theory, the safety boundary is not the only boundary beyond which (safe) work is impossible. In fact, systems have to be safe *and* economically viable *and* functional, all at the same time. This is why control problems occur and why they can be so difficult to handle. These different requirements are often, if not always, in competition with each other. Rasmussen's (1997) dynamic safety model suggested that safety-critical work is bounded on three sides:

- A boundary of *acceptable performance*. Beyond this boundary, safety problems will occur. The risk of adverse events goes down when system operations move further away from this boundary. This will simultaneously move operations closer to other boundaries, however, like economic sustainability.
- A *workload boundary*. Beyond this boundary, there is more work than people inside the system are capable of accomplishing.
- An *economic boundary*. Beyond this boundary, the system is economically no longer viable. It may be safe and not involve high workloads for those working inside it, but it will not be sustainable.

Together, these three boundaries form the safety envelope of the system inside of which work is at all possible, economically viable, and safe enough. All of these boundaries push on the work that goes on inside them, and forces from one side will soon be met by counterforces from another. Pressures to become more economic, for example, may quickly push work up against the workload and safety boundaries, and practitioners will start pushing back in various ways. In Rasmussen's (1997) words:

> The work space ... is bounded by administrative, functional and safety-related constraints. The normal changes found in local work conditions lead to frequent modifications of strategies and activity will show great variability. ... During the adaptive search the actors have ample opportunity to identify "an effort gradient" and management will normally supply an effective "cost gradient." The result will very likely be a systematic migration toward the boundary of functionally acceptable performance and, if crossing the boundary is irreversible, an error or an accident may occur. (p. 189)

Control theory does not see an organization as a static design of components or layers. It readily accepts that a system is more than the sum of its constituent elements. Instead, an organization is seen as a set of constantly changing and adaptive processes focused on achieving the multiple goals of the organization and adapting around its multiple constraints. The relevant units of analysis in control theory are therefore not components or their breakage (e.g., holes in layers of defense) but system constraints and objectives (N. Leveson, 2002; Rasmussen, 1997):

An important consequence is that control theory is not concerned with individual unsafe acts or errors or even individual events that may have helped trigger an adverse event. Such a focus does not help, after all, in identifying broader ways to protect the system against migrations toward risk. Control theory also rejects the depiction of adverse events in a traditionally physical way as the latent failure model does, for example. Accidents are not about particles, paths of traveling, or events of collision between hazard and process to be protected. Removing individual unsafe acts, errors, or singular events from an adverse event sequence only creates more space for new ones to appear if the same kinds of systemic constraints and objectives are left similarly ill controlled in the future. The focus of control theory is therefore not on erroneous actions or violations but on the mechanisms that help generate such behaviors at a higher level of functional abstraction—mechanisms that turn these behaviors into normal, acceptable, and even indispensable aspects of an actual, dynamic, daily work context that needs to survive inside the constraints of three kinds of boundaries (functional, economic, and safety).

For control theory, the making and enforcing of rules is not an effective strategy for controlling behavior. This, instead, can be achieved by making the boundaries of system performance explicit and known and to help people develop skills at coping with the edges of those boundaries. Ways proposed by Rasmussen (1997) include increasing the margin from normal operation to the safety boundary. This can be done by moving the safety boundary further out or by moving operations further inward, away from a fixed safety boundary. In both cases, more margin is open. This, however, is only partially effective because of risk homeostasis—the tendency for a system to gravitate back to a certain level of risk acceptance, even after interventions to make it safer. In other words, if the boundary of safe operations is moved further away, then normal operations will likely follow not long after—under pressure, as they always are, from the objectives of efficiency and less effort.

Leaving both pressures in place (a push for greater efficiency and a safety campaign pressing in the opposite direction) does little to help operational people cope with the actual dilemma at the boundary. Also, a reminder to try harder and watch out better, particularly during times of high workload, is a poor substitute for actually developing skills to cope at the boundary. Raising awareness, however, can be meaningful in the absence of other possibilities for safety intervention, even if the effects of such campaigns tend to wear off quickly. Greater safety returns can be expected only if something more fundamental changes in the behavior-shaping conditions or the particular process environment (e.g., the relief of a bed crunch in the emergency department after a busy Saturday night). In this sense, it is important to raise awareness about the migration toward boundaries throughout the organization at various managerial levels. A fuller range of countermeasures then becomes available beyond telling frontline operators to be more careful. Organizations that are able to do this effectively have sometimes been dubbed high-reliability organizations.

HIGH-RELIABILITY THEORY

So, what are the human actions, the deliberate processes by which risks are monitored, evaluated, and reduced in organizations? There is a school of thought that argues that careful organizational practices can make up for the inevitable limitations on the rationality of individual members. High-reliability theory describes the extent and nature of the effort in which people at all levels in an organization can engage to ensure consistently safe operations despite inherent complexity and risks (Marone & Woodhouse, 1986).

Through a series of empirical studies, high-reliability organizational (HRO) researchers found that through leadership safety objectives, the maintenance of relatively closed systems, functional decentralization, the creation of a safety culture, redundancy of equipment and personnel, and systematic learning, organizations could achieve the consistency and stability required to effect failure-free operations. Some of these categories were inspired by the worlds studied—naval aircraft carriers, for example. There, in a relatively self-contained and isolated, closed system, systematic learning was an automatic by-product of the swift rotations of naval personnel, turning everybody into instructor and trainee, often at the same time. Functional decentralization meant that complex activities (like landing an aircraft and arresting it with the wire at the correct tension) were decomposed into simpler and relatively homogeneous tasks, delegated down into small work groups with substantial autonomy to intervene and stop the entire process independent of rank. High-reliability researchers found many forms of redundancy—in technical systems, supplies, even decision-making and management hierarchies, the last through shadow units and gaining multiple skills.

When researchers first set out to examine how safety is created and maintained in such complex systems, they focused on errors and other negative indicators, such as incidents, assuming that these were the basic units that people in these organizations used to map the physical and dynamic safety properties of their production technologies, ultimately to control risk (Rochlin, 1999). The assumption was wrong: They were not. Operational people, those who work at the sharp end of an organization, hardly defined safety in terms of risk management or error avoidance. Four ingredients kept reappearing, and they form the contours of what has become high-reliability theory or the theory of high-reliability organizations.

Leadership safety objectives are the first ingredient. For a high-reliability organization in healthcare to blossom, hospital leadership needs to declare patient safety an urgent and top priority (Morath & Turnbull, 2005). Without such commitment, there is little point in trying to promote a culture of reliability. The idea is that others in the organization will never be enticed to find safety more important than their leadership. Short-term efficiency, or acute production goals, are openly (and sometime proudly) sacrificed when chronic safety concerns come into play. Agreement about the core mission of the organization (safety) is sought at every available opportunity, particularly through clear and consistent top-down communication about the importance of safety. Such commitments and communication may not be enough, of course. Other theorists have pointed out that the distance between loftily stated goals and real action is large, and can remain large, despite leadership or managerial pledges to the contrary (Sagan, 1993).

The need for redundancy is the second ingredient. The idea behind redundancy is that it is the only way to build a reliable system from unreliable parts. Multiple and independent channels of communication and double checks can, in theory, produce a highly reliable organization (although the perils of redundancy were noted in the section on normal accident theory). Redundancy in high-reliability theory can take two forms, duplication and overlap. In duplication, two different units, people, or parts perform the same function, often in real time. Duplication is also possible serially, as in the double checking of a medication preparation. Overlap is redundancy, by which units or people or parts have some functional areas in common but not all. It is obviously a cheaper solution (Rochlin, LaPorte, & Roberts, 1987).

Decentralization, culture, and continuity form the third ingredient. High-reliability organizations rely on considerable delegation and decentralization of decision authority about safety issues. These organizations do not readily court government or regulatory interference with their activities and instead acknowledge the superiority of local entrepreneurial efforts to improve safety through engineering, procedure, or training (LaPorte & Consolini, 1991). Active searching and exploration for ways to do things more safely is preferred over passively adapting to regulation or top-down control (Wildavsky, 1988). People inside organizations continually create safety through their evolving practice (Woods, Dekker, Cook, Johannesen, & Sarter, 2010).

As a result, sharp-end practitioners in high-reliability organizations are entrusted to take appropriate actions in tight situations because they will have been inculcated in reliability through rituals, values, exercises, and incentives (recall that the effectiveness of these processes of inculcation met with considerable skepticism from garbage can and normal accident theorists). Having members work in a "total institution," isolated from wider society and inside their own world, seems to contribute to a culture of reliability. This aim is consistent with the maintenance of a relatively closed system. Finally, continuous operations and training, nonstop on-the-job education, a regular throughput of new students or other learners, and challenging operational workloads contribute greatly to reduced error rates and enhanced reliability.

These features, which form the basis for a possible high-reliability organization, are not entirely foreign to healthcare. They are, for example, echoed in calls to preserve the institutionalized house officer or resident. That vocation involves a constant presence among patients and their possible (safety) problems, a challenging workload, an isolation from wider society, and an incessant exposure to the values, rituals, and incentives of working inside medicine.

Organizational learning is the fourth ingredient for high-reliability organizations. High reliability grows out of incremental learning through trial and error. Things are attempted, new procedures or routines are tested (if carefully so), the effects are duly considered. Smaller dangers are courted to understand and forestall larger ones (Wildavsky, 1988). Simulation and imagination (e.g., disaster exercises) are important ways of doing so when the costs of failure in the real system are too high. For high-reliability theory, such learning does not need to be centrally orchestrated. In fact, the distributed, local nature of learning is what helps new and better ways of doing things emerge. The appearance of medical simulator centers around the world, as well as locally produced low-fidelity simulation technologies for the training of laparoscopy, are examples of such bottom-up learning by trial and error.

Ensuing empirical work, stretching across decades and a multitude of high-hazard, complex domains (aviation, nuclear power, utility grid management, naval operations) affirmed this richer picture. Operational safety—how it is created, maintained, discussed, mythologized—is much more than the control of negatives. As Rochlin (1999) put it:

> The culture of safety that was observed is a dynamic, intersubjectively constructed belief in the possibility of continued operational safety, instantiated by experience with anticipation of events that could have led to serious errors, and complemented by the continuing expectation of future surprise. (p. 1549)

The creation of safety, in other words, involves a belief about the possibility to continue operating safely. This belief is built up and shared among those who do

the work every day. It is moderated or even held up in part by the constant preparation for future surprise—preparation for situations that may challenge people's
current assumptions about what makes their operation risky or safe. It is a belief
punctuated by encounters with risk, but it can become sluggish by overconfidence
in past results, blunted by organizational smothering of minority viewpoints, and
squelched by acute performance demands or production concerns. But, that also
makes it a belief that is, in principle, open to organizational or even regulatory
intervention to keep it curious, open minded, complexly sensitized, inviting of
doubt, and ambivalent toward the past (Weick, 1993).

Safety, however, is not the same as reliability. A part can be reliable, but in and
of itself it cannot be safe. It can perform its stated function to some expected level or
amount, but it is context, the context of other parts, of the dynamics, the interactions,
and the cross adaptations between parts, that makes things safe or unsafe. Reliability
as an engineering property is expressed as a failure rate of a component over a period
of time. In other words, it addresses the question of whether a component lives up
to its prespecified performance criteria. This, indeed, also is what quality improvement and control can be about. Quality and reliability are often associated with a
reduction in variability, and an increase in replicability: The same process, narrowly guarded, produces the same predictable outcomes. In that sense, the choice
of the word *reliability* may be slightly unfortunate for a system like healthcare. As
discussed, healthcare has a complex relationship with the promises and perils of
standardization.

KEY POINTS

- The major source of patient safety risk lies not with individual caregivers but with the system surrounding those caregivers—the organization,
  administration, design, resourcing, and technology of healthcare. This is
  where risk to patients brews and grows and where risk should be most effectively recognized, managed, and contained.
- The safety literature features the diversity of ideas about what risk is, how it
  builds at the blunt end of an organization, and how it can be controlled. This
  diversity means that there are a number of different ways in which healthcare can conceptualize its risk and engage in attempts to control it.
- Risk can be seen as energy whose accidental or inappropriate release needs to
  be contained. It can also be seen as a structural property of complex systems,
  as the gradual organizational acceptance of deviations, or as a managerial or
  control problem. This is how different traditions in the safety literature have
  conceptualized it, leading to different approaches and countermeasures.
- Thus, risk can be seen as something that is best managed by putting in extra
  barriers, by questioning the notion of "normal" operations, by reducing
  complexity, or by enhancing the control system and managerial or leadership commitments surrounding it. None of these approaches is necessarily
  privileged, and taking a combination of perspectives on risk can be the best
  investment in patient safety.

REFERENCES

Aiken, L. H., Clarke, S. P., Sloane, D. M., Sochalski, J., & Silber, J. H. (2002). Hospital nurse staffing and patient mortality, nurse burnout, and job dissatisfaction. *Journal of the American Medical Association, 288*(16), 1987–1993.

Berwick, D. M., Calkins, D. R., McCannon, C. J., & Hackbarth, A. D. (2006). The 100,000 lives campaign: Setting a goal and a deadline for improving health care quality. *Journal of the American Medical Association, 295*(3), 324–327.

Clarke, L., & Perrow, C. (1996). Prosaic organizational failure. *American Behavioral Scientist, 39*(8), 1040–1057.

Cohen, M. D., March, J. G., & Olsen, J. P. (1988). A garbage can model of organizational choice. In J. G. March (Ed.), *Decisions and organizations* (pp. 294–334). Oxford, UK: Blackwell.

Columbia Accident Investigation Board. (2003). *Report volume 1, August 2003.* Washington, DC: Author.

Cook, R., & Rasmussen, J. (2005). "Going solid": A model of system dynamics and consequences for patient safety. *Quality and Safety in Health Care, 14*(2), 130–134.

Cook, R. I., Render, M., & Woods, D. D. (2000). Gaps in the continuity of care and progress on patient safety. *British Medical Journal, 320*(7237), 791–795.

Dekker, S. W. A. (2003). Failure to adapt or adaptations that fail: Contrasting models on procedures and safety. *Applied Ergonomics, 34*(3), 233–238.

Dekker, S. W. A. (2007). Discontinuity and disaster: Gaps and the negotiation of culpability in medication delivery. *Journal of Law, Medicine and Ethics, 35*(3), 463–470.

Feynman, R. P. (1988). *"What do you care what other people think?": Further adventures of a curious character.* New York: Norton.

Hollnagel, E. (2004). *Barriers and accident prevention.* Aldershot, UK: Ashgate.

Hugh, T. B., & Dekker, S. W. A. (2008). Laparoscopic bile duct injury: Understanding the psychology and heuristics of the error. *ANZ Journal of Surgery, 78*(12), 1109–1114.

Jensen, C. (1996). *No downlink: A dramatic narrative about the* Challenger *accident and our time.* New York: Farrar, Straus, Giroux.

Langewiesche, W. (1998). *Inside the sky: A meditation on flight.* New York: Pantheon Books.

LaPorte, T. R., & Consolini, P., M. (1991). Working in practice but not in theory: Theoretical challenges of "high-reliability organizations." *Journal of Public Administration Research and Theory: J-PART, 1*(1), 19–48.

Leveson, N., Daouk, M., Dulac, N., & Marais, K. (2003). *Applying STAMP in accident analysis.* Cambridge, MA: Engineering Systems Division, Massachusetts Institute of Technology.

Leveson, N. G., & Turner, C. S. (1993). An investigation of the Therac-25 accidents. *Computer, 26*(7), 18–41.

Marone, J. G., & Woodhouse, E. J. (1986). *Averting catastrophe: Strategies for regulating risky technologies.* Berkeley: University of California Press.

Morath, J. M., & Turnbull, J. E. (2005). *To do no harm: Ensuring patient safety in health care organizations.* San Francisco: Jossey-Bass.

Ong, M., Bostrom, A., Vidyarthi, A., McCulloch, C., & Auerbach, A. (2007). House staff team workload and organization effects on patient outcomes in an academic general internal medicine inpatient service. *Archives of Internal Medicine, 167*, 47–52.

Paget, M. A. (2004). *The unity of mistakes: A phenomenological interpretation of medical work.* Philadelphia: Temple University Press.

Perrow, C. (1984). *Normal accidents: Living with high-risk technologies.* New York: Basic Books.

Pidgeon, N., & O'Leary, M. (2000). Man-made disasters: Why technology and organizations (sometimes) fail. *Safety Science, 34*(1–3), 15–30.

Pronovost, P. J., & Vohr, E. (2010). *Safe patients, smart hospitals.* New York: Hudson Street Press.

Rasmussen, J. (1997). Risk management in a dynamic society: A modelling problem. *Safety Science, 27*(2–3), 183–213.

Reason, J. T. (1990). *Human error*. New York: Cambridge University Press.

Rochlin, G. I. (1999). Safe operation as a social construct. *Ergonomics, 42*(11), 1549–1560.

Rochlin, G. I., LaPorte, T. R., & Roberts, K. H. (1987). The self-designing high reliability organization: Aircraft carrier flight operations at sea. *Naval War College Review*, 76–90.

Rosness, R., Guttormsen, G., Steiro, T., Tinmannsvik, R. K., & Herrera, I. A. (2004). *Organisational accidents and resilient organizations: Five perspectives (Revision 1)* (No. STF38 A 04403). Trondheim, Norway: SINTEF Industrial Management.

Sagan, S. D. (1993). *The limits of safety: Organizations, accidents, and nuclear weapons*. Princeton, NJ: Princeton University Press.

Snook, S. A. (2000). *Friendly fire: The accidental shootdown of U.S. Black Hawks over northern Iraq*. Princeton, NJ: Princeton University Press.

Starbuck, W. H., & Milliken, F. J. (1988). *Challenger*: Fine-tuning the odds until something breaks. *The Journal of Management Studies, 25*(4), 319–341.

Turner, B. A. (1978). *Man-made disasters*. London: Wykeham.

Vaughan, D. (1996). *The* Challenger *launch decision: Risky technology, culture, and deviance at NASA*. Chicago: University of Chicago Press.

Vaughan, D. (1999). The dark side of organizations: Mistake, misconduct, and disaster. *Annual Review of Sociology, 25*, 271–305.

Vaughan, D. (2005). System effects: On slippery slopes, repeating negative patterns, and learning from mistake? In W. H. Starbuck & M. Farjoun (Eds.), *Organization at the limit: Lessons from the* Columbia *disaster* (pp. 41–59). Malden, MA: Blackwell.

Wachter, R. M. (2008). *Understanding patient safety*. New York: McGraw-Hill.

Weick, K. E. (1990). The vulnerable system: An analysis of the Tenerife air disaster. *Journal of Management, 16*(3), 571–594.

Weick, K. E. (1993). The collapse of sensemaking in organizations: The Mann Gulch disaster. *Administrative Science Quarterly, 38*(4), 628–652.

Weingart, P. (1991). Large technical systems, real life experiments, and the legitimation trap of technology assessment: The contribution of science and technology to constituting risk perception. In T. R. LaPorte (Ed.), *Social responses to large technical systems: Control or anticipation* (pp. 8–9). Amsterdam: Kluwer.

Westrum, R. (1993). Cultures with requisite imagination. In J. A. Wise, V. D. Hopkin, & P. Stager (Eds.), *Verification and validation of complex systems: Human factors issues*. Berlin: Springer-Verlag.

Wildavsky, A. B. (1988). *Searching for safety*. New Brunswick, NJ: Transaction Books.

Woods, D. D., Dekker, S. W. A., Cook, R. I., Johannesen, L. J., & Sarter, N. B. (2010). *Behind human error*. Aldershot, UK: Ashgate.

Woods, D. D., & Hollnagel, E. (2006). *Joint cognitive systems: Patterns in cognitive systems engineering*. Boca Raton, FL: CRC/Taylor & Francis.

Woods, D. D., Patterson, E. S., & Cook, R. I. (2005). *Behind human error: Taming complexity to improve patient safety*. Columbus, OH: Institute for Ergonomics, The Ohio State University.

Wynne, B. (1988). Unruly technology: Practical rules, impractical discourses and public understanding. *Social Studies of Science, 18*(1), 147–167.

6 Practical Tools for Creating Safety

Learning about the limits and difficulties of improving patient safety is one thing, it is necessary. The previous chapters have discussed a host of issues—from the problems of the medical competence hierarchy to local rationality to the complexities and capabilities of new technology and the administrative and managerial origin of organizational adverse events. Knowing what to do with that knowledge can be something else. What to do now? This chapter discusses a number of human factors interventions that have proven worthwhile in both healthcare and other industries. It begins with safety reporting and organizational learning, then moves to adverse event investigations and resource management training, and finishes with a human factors consideration of checklists as tools for improving safety.

SAFETY REPORTING AND ORGANIZATIONAL LEARNING

WHAT IS ORGANIZATIONAL LEARNING?

The literature on organizational learning is diverse, but it seems to agree on one thing: Organizational learning is more than just the acquisition and enactment of new knowledge by individuals. In fact, as Argyris and Schön (1978) pointed out, most organizations know less than their members do, and in many cases organizations seem incapable of learning what their members know. This of course raises the question of who or what does the learning. Some contend that only individuals can learn. Yet their lessons can become institutionalized so that their impact survives over time and influences all kinds of organizational activities. Organizational learning, in this reading, is a process of structuration in which individuals learn and enact structures, which in turn create opportunities and constraints for further action and learning by others (Mahler, 2009). This can even become embedded in the stories, myths, values, and cultural assumptions that an organization holds about itself (Rochlin, 1999; Schein, 1992). Organizational learning as a topic has long intrigued managers and theorists alike:

> In essence, organizational learning is about the astuteness of the organization, and the honesty and curiosity of its members in uncovering problems. It focuses on the capacity of organization members to assess causes and effects … and to search out and reflect on possible solutions, to make choices, and to incorporate them into the organizational establishment. (Mahler, 2009, p. 25)

Is learning different from other forms of organizational change? For sure, in the wake of an adverse event, regulators or bad publicity may force an organization to make changes (get rid of people, put a reminder in place, develop a new guideline or policy) that cannot be construed as learning (Cook & Nemeth, 2010) but rather as a strategic adaptation or a tactical bow to political or other force fields. In fact, such changes may mean that the organization is actually learning the wrong thing (Vaughan, 1999). One could, however, also argue that any change is equal to learning (even if it means learning to cope with political or media pressure). But learning is generally seen as change that is directional, as something that is consistent with the core values and objectives (beyond sheer survival) of the organization.

Learning requires more than obtaining new information about things. A lot must happen, or organizations will not learn at all (Mahler, 2009):

> The information needed to recognize problems or search out solutions may not be available. Many organizational actors ignore internal or external warning signals of unsatisfactory results because they do not think changes are possible, they do not want to admit that their performance is unsatisfactory, or they do not want to undertake unpopular or troublesome change. Officials may not know how to use experience or information about negative results to "fix" programs. (p. 19)

For the purpose of this chapter, we can see organizational learning as a collection of processes with which organizations improve their ability to accomplish their objectives by analyzing their past efforts (Mahler, 2009). Incident reporting and investigating are obviously a good way of doing this, although of course Mahler's reflections indicated strongly that other things need to happen as well. People need to know where to find the sort of information from which they can. They need to be able to extract meaningful lessons from it and convert those into structures and processes that can survive as organizational memory, as constraints and opportunities for future action inside the organization. Just reporting bad news is only a start, if a good start.

EFFECTIVE REPORTING SYSTEMS: NONPUNITIVE, PROTECTED, VOLUNTARY

A number of industries have implemented and refined reporting systems. They are alive and well in aviation, in nuclear power generation, in many military applications, in petrochemical processing, even in steel production (American Medical Association [AMA], 1998; Billings, 1997). In healthcare, reporting was one of the main agenda points of the patient safety report by the Institute of Medicine (Kohn, Corrigan, & Donaldson, 2000). Near-miss reporting systems that work do a couple of things really well. They are nonpunitive, they are protected, and they are voluntary (Barach & Small, 2000).

Let us deal with the voluntary part first. Implementing an event reporting and investigation program raises the question of which events you want people to report and investigate. The point of such a program, of course, is to contribute to organizational learning and help prevent recurrence through systemic changes that aim to redress some of the basic circumstances in which caregiving work went awry. So, any event that has the potential to elucidate and improve the conditions for safe practice is, in principle, worth reporting and investigating. But that does not create intelligible guidance for healthcare workers or investigators.

What counts as a clear opportunity for organizational learning for one person, perhaps constitutes a dull event not worth reporting to somebody else. Something that could have gone terribly wrong, but did not, can also produce this interpretational ambiguity. After all, in medicine, particularly emergency medicine, things can go terribly wrong all the time, but that does not make reporting everything particularly meaningful.

Which event is worthy of reporting and investigating is, at its heart, a judgment. First, it is a judgment by those who perform safety-critical work at the sharp end. This judgment is shaped foremost by experience—the ability to deploy years of practice into gauging the reasons and seriousness behind a misalignment of expectation and outcome. To be sure, the same experience can have a way of blunting the judgment of what to report. If all has been seen before, why still report? What individuals and groups define as "normal" can glide, incorporating more and more nonconformity as time goes by and experience accrues. In addition, the rhetoric used to talk about mistakes in medicine can serve to "normalize" (or at least deflect) an event away from the caregivers at that moment. A "complication" or "noncompliant patient" is not so compelling to report (although perhaps worth sharing with peers in some other forum), as when the same event were to be denoted as, for example, a diagnostic error. Whether an event is worth reporting, in other words, can depend on what language is used to describe that event in the first instance.

Social systems, in other words, are expert at adapting their readings of risk to accommodate the seemingly normal and leave intervention only for the patently hazardous. Norms for what counts as risky are renegotiated the whole time, particularly as operational experience with a particular system, particular procedure, or particular failure mode (or disease) accumulates. It is the kind of normalization of deviance ("Oh, we've seen this before, it's okay.") that eventually brings space shuttles down in flames (Jensen, 1996; Mahler, 2009; Rogers et al., 1986; Vaughan, 1996). If it is okay, it is not a near miss. It will remain unknown as risky to everybody else.

It is not strange that underreporting of adverse events in healthcare has been estimated as between 50% and 96% (Kohn et al., 2000; Leape, 1994), even though this assumes a "norm" that can be agreed on about what constitutes an event relative to which people underreport. Instead of adverse event, sentinel event, or incident, some organizations apply the label of a "near miss." In many hospitals, a near miss could be considered much more a "miss" than anything "near." If it is a miss, it is a nonevent. And if it is a nonevent, there is nothing to report. What is the standard procedure at the hospital? Wipe the brow, say, "Whew, we got away with that one," then peel off the gloves and go home—no harm, no incident; no near miss, no report. No signal is seen among the noise. This has another interesting implication: In some cases, a lack of experience (either because of a lack of seniority or because of inexperience with a particular case or in that particular department) can be immensely refreshing in questioning what is normal (and thus what should be reported or not).

This does attest to the notion that having a mandatory system makes no sense. If reporting a near miss is mandatory, then practitioners will likely engage in various kinds of rhetoric or interpretive work to decide that the event they were involved in was not a near miss. Of course, this does not mean that an organization cannot in some way try to characterize what it finds important. Investing in a meeting at which different department stakeholders share their examples of what is worth reporting could be worthwhile. It could result in a list of examples that can be handed to people as partial guidance on what to report. In the end, given the uncertainties about how things can be seen as valuable by other people and how they could have developed to perhaps produce harm, the ethical obligation should be: "If in doubt, report." But that still cannot make things obligatory. Practitioners may simply say that they never were in doubt. Indeed, having a mandatory system would probably increase the underreporting rate. And making reporting mandatory implies some kind of sanction if something is not reported, which destroys the first ingredient for success: having a nonpunitive system.

Nonpunitive means that the reporter is not punished for revealing personal violations or other breaches or problems of conduct that might be construed as culpable. This is normally seen as hugely problematic (see also Chapter 7). The infallibility hypothesis (see Chapter 1) seems to leave few alternatives other than seeing an error as something bad, shameful, undesirable, or even negligent. Near misses that cannot be constructed or rationalized away as remediable, irredeemable incompetence, or bad luck (and thus might actually be reported) have to be due to some potentially culpable omission or act.

But not punishing that which is reported makes great sense, if anything because otherwise it will not get reported. One voluntary report about an adverse medication event in a Swedish hospital, for example, ended up in the media. The nurse who had reported it was identified, found, charged with a crime, and convicted (Dekker, 2007). The willingness of another nurse to report anything would have taken a severe beating.

Then, finally, successful reporting systems are *protected,* which means that reports are confidential rather than anonymous. What is the difference? *Anonymity* typically means that the reporter is never known, not to anybody. No name or affiliation has to be filled in anywhere. *Confidentiality* means that the reporter fills in name and affiliation and is thus known to whomever receives the report. But from that point, the identity is protected under any variety of industrial, organizational, or legal arrangements.

If reporting is anonymous rather than confidential, two things might happen quickly. The first is that the reporting system becomes the garbage can for any vitriol that practitioners may accumulate about their job, their colleagues, or their hospital during a workday, workweek, or career. The risk for this, of course, is larger when there are few meaningful or effective line management structures in place that could take care of such concerns and complaints. Senseless and useless bickering could then clog the pipeline of safety-critical information. Signals of potential danger would get lost in the noise of potent grumble. So, that is why confidentiality makes more sense. The reporter may feel some visibility, some accountability even, for reporting things that can help the organization learn and grow.

The second problem with an anonymous reporting system is that the reporter cannot be contacted if the need arises for any clarifications. The reporter is also out of reach for any direct feedback about actions taken in response to the report. The

NASA (National Aeronautics and Space Administration) Aviation Safety Reporting System (ASRS) recognized this quickly after finding reports that were incomplete or could have been much more potent in revealing possible danger if only this or that detail could be clarified. As soon as a report is received, the narrative is separated from any identifying information (about the reporter and the place and time of the incident) so that the story can start to live its own life without the liability of recognition and sanction appended to it. This recipe has been hugely successful. ASRS receives more than 1,000 reports a week (Billings, 1997).

This does raise the question of who can know about a near miss that is reported. How can people trust that their reports will not be used against them? What kinds of protections are fair? And will such protections not lead to the abuse of the near-miss reporting system for accountability (rather than learning) purposes it was never intended to support (like practitioners submitting a report because it provides them a level of protection, not because they want to help the organization or their peers learn)?

This, indeed, is a risk inherent in setting up a reporting system. For some cases, the reporting system can become a liability management device more than an instrument for collective learning and organizational safety improvement. The system does not become much wiser, but practitioners may become better protection against any consequences of their actions. This will likely be seen as unfair by other members of society.

NARRATIVES OR INDEXING?

So, what delimits an "event"? A reporter needs to decide the beginning and end of the reported event. This reporter needs to decide how to describe the roles and actions of other participants who contributed to the event (and to what extent to identify other participants, if at all). Finally, the reporter needs to settle on a level of descriptive resolution that offers the organization a chance to understand the event and find leverage for change. Many of these things can be structured beforehand, for example, by offering a reporting form that gives guidance and asks particular questions ("need to know" for the organization to make any sense of the event) as well as ample space for free-text description.

Free-text narratives take time to write and to read, but they are often indispensable for understanding what happened and for finding levers to do something about it. Many reporting systems, however, use indexing systems. These are not necessarily automatic but rather are tools used by staff to chop the narrative up into categories. They allow information from the near-miss database to be presented numerically and graphically and be compared across space or time (as in "we have so many of these so-and-so incidents with this or that technology compared with only so many last month or compared to only so many in our sister hospital across town"). Often, managers see nothing *but* such bar charts or pie charts or number tables in the hope that it would present them with some actionable information.

It probably does not, at least not often. The whole point of the narrative *is* the narrative. Outside the story, there is no near miss, there is no incident. There is no buildup, no context, no resolution, only dead remnants arbitrated by somebody who was not there when it happened, classified remains that have been lobotomized out of the living story from which they came. This limitation, and the idea that classification should not be confused with analysis, was noted by the AMA in 1998 (p. 41), at a time when initial enthusiasm for reporting was building in healthcare (Kohn et al., 2000):

> Classification does involve a type of analysis but a type that greatly constrains the insights that can be obtained from the data. Typically, when classification systems are used as the analysis, a report of an incident is assigned, through a procedure or set of criteria, into one or another fixed category. The category set is thought to capture or exhaust all of the relevant aspects of failures. Once the report is classified the narrative is lost or downplayed. Instead, tabulations are built up and put into statistical comparisons. Put simply, once assigned to a single category, one event is precisely, and indistinguishably like all the others in that category. Yet research on human performance in incidents and accidents emphasizes the diversity of issues and interconnections. ... Capturing a rich narrative of the sequence and factors involved in the case has proven essential. Often, new knowledge or changing conditions leads investigators to ask new questions of the database of narratives. The analyst often goes back to the narrative level to look for new patterns or connections. (AMA, 1998, p. 41)

Categories force analysts to make decisions, to draw lines, to decide that an action or circumstance is *this* but not *that*. What if it is both? Or neither? Attempts to map situated human capabilities such as decision making, proficiency, or deliberation onto discrete categories are doomed to be misleading for they cannot cope with the complexity of actual practice without serious degeneration (Angell & Straub, 1999). Classification disembodies data. It removes the context that helped produce the near miss in its particular manifestation. This disables understanding because by excising performance fragments away from their context, classification destroys the local rationality principle. This has been the fundamental concept for understanding—not judging—human performance for the last 50 years: People's behavior is rational, if possibly erroneous, when viewed from the inside of their situations, not from the outside and from hindsight.

Remember the local rationality principle from Chapter 2. It reminds us that consequences of actions are not necessarily well correlated with intentions, yet this evaporates in the wake of near-miss classification. The point in learning about near misses is not to find out where people went wrong. It is to find out why their assessments and actions made sense to them at the time given their knowledge, goals, tools, and resources. We have to assume that if it made sense to someone (given the background and circumstances), it also will make sense to someone else, and the near miss will not only repeat itself but also perhaps not remain a near miss next time. Controversial behavior can be made to make sense (read: understood) once resituated in the context that brought it forth.

After an observation of a near miss is tidily locked away into some category, it has been objectified, formalized away from its context. Without context, there is no way to reestablish local rationality. And without local rationality, there is no way to understand human error. Classification probably presents managers, administrators, and possibly even practitioners an illusion of understanding. It disconnects human agents' performance from the context that brought it forth and from the circumstances that accompanied it, that gave it meaning, and that hold the keys to its explanation.

We need narratives about near misses, which can be as much stories of how things went wrong as stories about how things (which could have gone wrong) went right after all. They can be stories of resilience, in other words (see Chapter 8). Stories of how a system of people and technologies was able to recognize, absorb, and adapt to changes and challenges that perhaps fell outside what the team was trained or designed to handle (Hollnagel, Woods, & Leveson, 2006).

Nonetheless, the idea that near-miss reporting is simply about counting negatives, presented to management and administrations in the form of various graphics, may be compelling to the healthcare industry for the same reasons that any numerical performance measurement is. If anything, the reasons for such quantitative reliance may be highly practical. The interface that safety departments or governance units in hospitals have with senior leadership often consists only of a communication of numbers—so much of this kind or that over the past month in a little table of categories with hit rates next to them, for example. There may not be time to go into stories, to go into depth. And individual stories do not show directions, trends, trajectories—the sorts of things that managers want to know because they believe that they have control (or need to show that they can exert control) over such things.

The quantification of adverse event results creates funny side effects that uphold and sponsor this belief. Showing numbers helps assert the existence of an observer-independent reality. The quantified results are an objective window into this reality that can be offered to managers, with the illusion that here is a world that they can affect, influence, mold, shape, control, and understand without having to take some analyst's word for it (Gergen, 1999):

> The use of figures and graphs not only embodies numbers, but gives the reader the sense of "seeing the phenomenon." By using figures and graphs the scientist implicitly says, "You don't have to take my word for it, look for your self." (p. 56)

What is finessed here is what the numbers "stand for" at the beginning and how they were derived. Confidence about where to intervene is reduced to a kind of ranking or numerical strength, a sort of democracy of numbers. The problem, of course, is that managers still *do* take the analyst's word for it. It was the analyst, after all, who decided to put some things into some categories and others into other categories. This, however, as well as the rationale for it, has disappeared entirely by the time the numbers show up in the boardroom. Collection and categorization are a lot easier than analysis. Analysis is easier than taking meaningful action. And taking meaningful action is still miles removed from preventing an adverse event.

Near-miss classifications easily become the stand-in for analysis, real understanding, and actionable intelligence. Classification alone can become seen as a sufficient quantitative basis for managerial interventions. Data from the operation that have been excised and formalized away from their origin can be converted into graphs and bar charts, which are subsequently engineered into interventions. Never mind that the bar charts show comparisons between apples and oranges (e.g., causes and consequences of a near miss) that lead managers to believe that they have learned something of value. It may not matter because managers can elaborate their idea of control over operational practice and its outcomes.

The hospital world of real people and processes, of course, is not so easily fooled: Managerial "control" exists only in the sense of purposefully formulating and trying to influence healthcare workers' intentions and actions (Angell & Straub, 1999). It is not at all the same as being in control of the consequences (by which safety ultimately is measured healthcare-wide)—the real world is too complex and operational environments too stochastic. Numbers that may seem managerially appealing are really sterile, inert. They do not reflect any of the nuances of what it is to "be there," doing the work, creating safety on the line. Yet this is what ultimately determines safety (as outcome): People's local actions and assessments are shaped by their own perspectives; they are self-referential, embedded in histories, rituals, interactions, beliefs, and myths, both of their organization and themselves as individuals.

GETTING PEOPLE TO REPORT AND KEEPING UP THE REPORTING RATE

Getting people to report is difficult. Keeping up the reporting rate once the system is running can be equally difficult, although often for different reasons. Getting people to report is about two major things: accessibility and anxiety. The means for reporting must be accessible (e.g., forms easily and ubiquitously available, not cumbersome). Anxiety can initially be significant. What will happen to the report? Who else will see it? Do I jeopardize myself, my career, my colleagues? Does this make legal action against me easier?

Getting people to report is about building trust: trust that the information provided in good faith will not be used against those who reported it. Such trust must be built in various ways. An important way is by structural (legal) arrangement. Making sure people have knowledge about the organizational and legal arrangements surrounding reporting is important: Disinclination to report is often related more to uncertainty about what *can* happen with a report than by any real fear about what *will* happen.

One organization, for example, has handed out little credit-card-size cards to its employees to inform them about their rights and duties concerning an incident. Another way to build trust is by historical precedent: making sure there is a good record for people to lean on when considering whether to report an event. But trust is hard to build and easy to break; one organizational or legal response to a reported event that shows that divulged information can somehow be used against the reporter can destroy months or years of building goodwill.

Keeping up the reporting rate is also about trust but in greater part about involvement, participation, and empowerment. Many people come to work with a genuine concern for the safety and quality of their professional practice. If, through reporting, they can be given an opportunity actually to contribute to visible improvements, then few other motivations or exhortations to report are necessary.

Making a reporter part of the change process can be a good way forward, but this implies that the reporter wants (dares) to be identified as such, and that managerial hierarchies have no qualms about taking onboard the humblest of employees in the search for improved safety and quality. Sending feedback to the department about any changes that result from reporting is also a good strategy, but it should not become the stand-in for doing anything else with reports.

Many organizations are enraptured in the belief that reporting is enough of a virtue in itself: If only people report errors and their self-confessions are distributed back to the operational community, then things will automatically improve, and people will feel motivated to keep reporting. This does not work for long. Active engagement with that which is reported is necessary. Active, demonstrable intervention on the back of reported information also is.

In many organizations, the line manager is the recipient of reports. This makes (some) sense: The line manager probably has responsibility for safety and quality in the primary processes and should have the latest information on what is or is not going well. But this practice has some side effects. One is that it hardly renders reporters anonymous (given the typical size of a department), even if no name is attached to the report. The other is that reporting can have immediate line consequences (an unhappy manager, consequences for one's chances to progress in a career). In fact, especially when the line manager himself or herself is part of the problem the reporter wishes to identify, such reporting arrangements all but stop the flow of useful information.

While keeping a line-reporting mechanism in place can be productive for the continuous improvement work of a department, consideration should be given to a separate, parallel confidential reporting system. The reports in this system should go to a staff officer (e.g., a safety or quality official), not a line manager, who has no stakes in the running of the department. The difference between that which is reported to a line manager and that which is written in confidential reports can be significant. Both offer several kinds of leverage for change, and not mining both data sources for improvement information is a waste for any organization.

The collection of near-miss reports is only the starting point. The enthusiasm with which we encourage people to report is seldom matched by our ability to do anything meaningful *with* the reports—so that people will stay encouraged to report in the future. The number of reports collected, or the increase in such numbers, is often seen as a reason to celebrate the success of the system. But this is not the success. Near-miss reporting is only a means to an end, not the end itself.

One of the most important ways in which near-miss narratives, in all their cultural, historical, and mythical richness, can contribute to learning by other people closely involved with the safety-critical processes is by integrating them into recurrent training. This is what happens often in the aviation industry. Mandatory yearly crew resource management training (see this chapter) that focuses on the soft skills (communication, interaction) for producing safe, successful outcomes can benefit hugely from the review of and critical reflection on cases from people's own operational environment. That is one place where the loop from near-miss report to organizational learning can be closed—perhaps more effectively so than with a pie chart beamed up on the wall during a management meeting.

ROOT CAUSES AND RESPONSIBILITIES

Many investigation methods today aim to help you find the "root causes" of an event. Errors are seen as symptoms or effects of deeper trouble rather than as causes of that trouble. But a root cause is never something we find; it is something we construct. What we end up seeing as causal (or where the method we use directs our attention) is a function of the questions we ask and the answers we set out to find. Root cause analyses (RCAs) are useful for that purpose: getting different people around the table to talk about an event, to ask questions about it. Without a structured analysis, these people might never have sat together in the first place.

That said, agreement must still be reached on what counts as the root causes of a particular event. Root causes are the deepest, but still changeable, factors that underlie an event. To identify "gravity" as a root cause can be technically correct (and even more of a root cause than anything less fundamental) but hardly useful from a change perspective. At the same time, this means that the process of identifying root causes is a process in identifying "doable" projects and is thus open to politics, resource battles, and even the agendas or pet peeves of particular stakeholders. The identification of a root cause can never be completely arbitrated by a method; the preconceptions, attentions, and interpretations of stakeholders involved in the investigation always enter somehow.

This also raises the issue of responsibility. If particular factors have been identified as causal, then who should bear responsibility for this? This question splits into two—a useful part and a counterproductive one. Let us take the latter first. Asking who is responsible for a contributory or causal factor in a retrospective, blame-seeking sense, is counterproductive to safety and quality. People will likely deploy defense mechanisms and duck rather than embrace their responsibility. It may even dissuade reporters from sending information in the future if they see that reports become a vehicle for the assignment of blame rather than a mechanism for learning. Asking who is responsible in a forward-looking sense, in contrast, is useful. It is intimately connected to the sorts of recommendations that should follow from an investigation.

Recommendations for change should be sufficiently challenging to give people the motivation to do some work for them, but at the same time, they should be doable, the outcome should be somehow visible or demonstrable, and somebody or some party should be assigned responsibility for implementing the recommendation and providing follow-up within a particular timeline.

HOW SHOULD YOUR SAFETY DEPARTMENT LOOK?

It was suggested in the discussion on incident reporting that a staff-based safety department—outside the managerial line—may be better positioned to help the organization learn. Locating the safety department in an organizational staff position, not somewhere down the line, should render it as independent as possible of (yet not insensitive to) contemporary economic and political concerns. Granted, it is important that a strong connection exists to line management; otherwise, even the most astute analyses of adverse events will not generate any organizational change and learning. And if necessary, a safety department should have direct access to relevant organizational decision-making levels without having to pass through various levels of non-practice-oriented middle management.

But are there any other characteristics that make a hospital safety department more effective? There are (Woods, 2006). The first is that a safety department needs to have access to significant and independent resources (both human and monetary). The cost of adverse events, to most hospitals, is considerable. The cost of even one operation that may need to be redone (quite apart from any damages that need to be paid) could be sufficient for the annual salary of one safety department analyst. The resources given to a safety department must be independent of production or financial fluctuations for two reasons.

First, safety monitoring and vigilance may actually have to go up when economic cycles go down or budgetary pressures go up. The department will probably need the resources most when the hospital can least afford them. This is when production and other pressures may become apparent as safety problems, for example, in staff or equipment shortages. Second, a safety department may contain practitioners who do safety work on a part-time basis (in fact, this is a good idea; further discussion is provided in this chapter). These practitioners have to somehow be shielded from production ebbs and flows because safety concerns often accelerate when production pressures go up (production pressures that could simultaneously rob the safety department of its people).

As commented on in the discussion on safety reporting, a safety department needs to secure a constructive involvement in management activities and decisions that affect trade-offs across safety and efficiency, as well as involvement in managerial actions in the wake of incidents. This could mean an end to weekly, monthly, or quarterly targets. Safety is not something that the safety department produces, so targets make little sense. Targets can perversely become goals in themselves, meaning that some events can be reclassified (e.g., as a less-severe adverse event) just to make the numbers look better. However common it may be, this is both unethical and self-destructive. Targets also have a way of favoring quantitative representations (bar charts) of supposed safety issues rather than qualitative intelligence on what is going on—when it is going on (not because the week or month is up).

People who do safety work in a hospital or other healthcare organization should endeavor to have continued grounding in operational reality. Having only full-time safety people can make a department less effective as these people can lose their idea (or never had one) of what it is to operate at the sharp end and what the shifting sources of risk out there may be. It is also important to avoid having one professional group (e.g., anesthesiologists or nurses) dominate the safety department. Otherwise, this will skew what is seen as safety data, as risk, and as meaningful (or politically acceptable) ways of doing something about it.

Of course, just being a practitioner does not qualify people to be members of a safety department. Staff members in safety management should have some (and recurrent) education in incident/accident investigation, human factors, report writing, presentation, and so forth. Without this, a safety department can quickly be filled with happy amateurs whose well-meaning efforts contribute little to the quality and credibility of the safety function.

What should the safety department generate in return? First, it should remain sensitive to legitimate organizational concerns about production pressures, reputation, and economic constraints. Showing that one cares only about safety can quickly make a department irrelevant as no organization exists just to be safe. It is there to attain other goals, and through sheer economic linkage, a safety department is an inextricable part of that.

Again, the major contribution to the learning of a hospital is the production of real, actionable safety intelligence, as opposed to distant tabulations of statistical data. This means that the output of a safety department has to show how safety margins may have been eroding over time and where the gap lies between what causes problems in the operation and what management believes causes problems. This will require the safety department to go beyond statistics and come up with persuasive intelligence. For example, it may have to lay bare the innards of an incident that can reveal to management how it could have been misinterpreting the sources of operational risk and its role in shaping that risk.

Woods (2006) summarized the attributes of an effective safety department as the four "I's": It is informed, independent, informative, and involved. A safety department is informed about how the organization is actually operating (as opposed to just taking distant numerical measurements). This requires the safety department to have its feet firmly planted in the operation and its ear to the ground (e.g., via confidential reporting).

A safety department functions as an independent voice that can challenge accepted interpretations of risk (e.g., a "safety attitude problem" perceived by management may actually be the expression of an entirely different problem deeper inside the organization). Having a strong voice at the top is not enough; the safety department must be tightly connected to actual work processes and their changing, evolving sources of risk.

The safety department needs to be informative in terms of helping reframe managerial interpretations of safety threats and helping management direct investments in safety when and where necessary. It is important not only that the safety department has access to relevant managerial decision-making channels, but also that it finds new ways to transmit its insights. In-depth analysis of critical incidents and meaningful discussion of this during meetings with ample time and where all relevant stakeholders are present could be effective in recalibrating how an organization looks at itself.

Finally, the department has to be involved constructively in organizational actions that affect safety. Activities such as responding to incidents, or granting waivers, should not be done without consulting the safety department because all of these activities can substantially influence downstream operational safety.

Of course, staying independent but involved is difficult; involvement may compromise independence and vice versa. Staying informed ("What is going on?") but informative ("How can we look with a fresh perspective on what is going on?") also conflicts. Learning about safety requires close contact with the risk inherent in operational practice, as well as distance for reflection. Fully resolving these conflicts is impossible, but managing them consciously and effectively is feasible. It requires that the safety department is large enough to carry various perspectives, that those in it know it has to be seen as a constructive participant in the organization's activities and decisions that affect the balance across production and safety goals, and that it stays firmly anchored in the details of daily practice to know what is going on there.

ADVERSE EVENT INVESTIGATIONS

Once a reporting system is in place, one question that often surfaces is whether to investigate a reported event further and who should do this. This can quickly become a matter of cost versus (possible) benefit. Because of resource limitations, hospitals (or their safety departments) need to be picky about what to investigate to any depth. This is true not only for adverse events that become known through the reporting system but also those that come to the attention of people by any other means.

Various organizations have set up event assessment teams that meet at regular intervals. They consist of various insider experts who consider and rank the possible risks and vulnerabilities associated with a reported event and decide what should be taken on for deeper investigation. The criteria that these experts use are often surprisingly heuristic (founded in history, hunch, tradition, or hobbyhorses, for example), despite many efforts to formalize risk assessment. For example, if a particular problem seems to repeat itself (even if in different guises), that may be good reason to investigate one of those events further to see which systemic factors underlie their repeated production.

The well-known format by which adverse events are often investigated (if they are) in hospitals is RCA or root cause analysis. These are normally conducted by management or administrators (or clinical governance staff), by people who have some training and experience to conduct such an analysis. No analysis method, of course, leads to the truth of what happened in an adverse event. By making certain assumptions (e.g., root causes exist and are meaningful to find), the analysis asks certain questions and invites certain answers. This may leave some other things unexplored. An RCA is usually an intense process that involves multiple people and may take hours, days, or even longer.

In this section on adverse event investigations, RCAs are not to be relied on as the blueprint for how to do this in hospitals or other healthcare settings. Instead, the basic idea of a sequence of events (itself only one way of modeling an adverse event; see Chapter 5) is used to illustrate how a human factors approach would take the steps from contextual (e.g., clinical) data to human performance interpretations of what happened. Achieving context-independent descriptions of adverse events is critical to pull out the systemic factors that went into creating them and learning lessons that can be shared and distributed across other contexts as well.

First and Second Stories

From the perspective of human factors research and practice over the past 60 years, a distinction could be made between first and second stories of human error (AMA, 1998; Dekker, 2002). While any adverse event could, in principle, be told from an infinite number of angles and perspectives (thus generating countless stories), the first and second story in some sense represent the bookends. The contrast between them pulls views on human error, its role, reality, and hopes about it being remediable, maximally apart. The distance between the first and the second story is therefore instructive.

In the first story, human error is seen as the cause of the adverse event. The organizations in which, or the engineered systems with which people work, are assumed to have been basically safe; their success is intrinsic if only people follow the appropriate rules and guidelines. The chief threat to safety comes from the inherent unreliability of people. In the first story, progress in safety can be made by protecting these systems from unreliable humans through selection, proceduralization, automation, training, and discipline.

The second story sees human error not as a cause but as a symptom of failure. Human error is a symptom of trouble deeper inside the system. Safety is not inherent in systems. The systems themselves are contradictions between multiple goals that people must pursue simultaneously. People have to create safety. Also, human error is not an unsystematic, sudden hiccup. It is systematically connected to features of people's tools, tasks, and operating environment. The way things are organized, designed, staffed, and presented has immediate and traceable connections to how people function. This has been the major theme in many of the previous chapters (particularly Chapters 3 and 4). Progress on safety comes from understanding and influencing these connections.

The second story of human error represents a substantial movement across the fields of human factors and organizational safety and encourages the investigation of factors that easily disappear behind the label "human error" (e.g., longstanding organizational deficiencies, design problems, and procedural shortcomings). The rationale is that human error is not an explanation for failure but instead demands an explanation, and that effective countermeasures start not with individual human beings who themselves were at the receiving end of much latent trouble but rather with the error-producing conditions present in their working environment.

Willingness to embrace the second story of human error is not always matched by ability to do so. When confronted by failure, it is easy to retreat into the old view: seeking out the "bad apples" and assuming that with them gone or reminded of applicable rules or guidelines, the system will be safer than previously. The emphasis of an investigation on proximal causes ensures that the adverse event remains the result of a few uncharacteristically ill-performing individuals who are not representative of the system or the larger practitioner population in it. It leaves existing beliefs about the basic safety of the system intact.

One reason for these reversions to first stories in healthcare is that there are no independent investigations of significant medical events that result in a public report. Every adverse event investigation is carried out by stakeholders. The result is that there is no regular supply of reliable, authoritative, scientifically grounded investigations of medical mishaps. Stakeholders control the information at every stage and do so in ways that make their pronouncements about the underlying events conform to economic or political realities surrounding their position (Cook, Nemeth, & Dekker, 2008).

Faced with a bad, surprising event, people seem more willing to change the individuals in the event, along with their reputations, rather than amend their basic beliefs about the system that made the event possible. Certainly, reconstructing the human contribution to an adverse event is not easy. Investigators are seldom—if ever—there when events unfolded around the people under investigation. As a result, their actions and assessments may appear not only controversial but also truly befuddling when seen from a different point of view. To understand why people did what they did, it is necessary to go back, to triangulate, and to interpolate, from a wide variety of sources, the kinds of mindsets that they had at the time. However, as described in Chapter 2, hindsight introduces bias, which, along with the multiple pressures and constraints that operate on almost every investigation (political as well as practical), works against investigations. First stories, rather than second stories, are often the outcome.

Managing the Hindsight Bias

Remember the hindsight bias from Chapter 2. One of the safest bets investigators or outside observers can make is that they know more about the adverse event than the people who were caught up in it—thanks to hindsight and outcome knowledge. Hindsight and outcome knowledge mean being able to look back, from the outside, on a sequence of events that led to an outcome that has already happened. It means that in hindsight, people have almost unlimited access to the nature of the situation that surrounded people at the time (where they actually were vs. where they thought they were; what state their system was in vs. what they thought it was in). Hindsight allows investigators to pinpoint what people missed and should not have missed, what they did not do but should have done.

From the perspective of the outside and hindsight (typically the investigator's perspective), the entire sequence of events is exposed—the triggering conditions, its various twists and turns, the outcome, and the true nature of circumstances surrounding the route to trouble. This contrasts fundamentally with the point of view of people who were inside the situation as it unfolded around them. To them, neither the outcome nor the entirety of surrounding circumstances was known. They contributed to the direction of the sequence of events on the basis of what they saw and understood to be the case on the inside of the evolving situation. For investigators, however, it is difficult to attain this perspective. The mechanisms by which hindsight operates on human performance data are mutually reinforcing. Together, they continually pull the investigators in the direction of the position of the retrospective outsider. The ways in which human performance evidence from the rubble of an adverse event are retrieved, represented, and retold typically sponsor this migration of viewpoint.

Tracing the sequence of events back from the outcome, which investigators already know, they invariably come across circumstances for which people had opportunities to revise their assessment of the situation but failed to do so; people were given the option to recover from their route to trouble but did not take it. These are counterfactuals—common in the analysis of adverse events. Counterfactuals prove what could have happened if certain minute and often utopian conditions had been met. Counterfactual reasoning may be a fruitful exercise when trying to uncover potential countermeasures against such failures in the future.

However, saying what people could have done to prevent a particular outcome does not explain why they did what they did. This is the problem with counterfactuals. When they are enlisted as explanatory proxy, they help circumvent the hard problem of investigations: finding out why people did what they did. Stressing what was not done (but if it had been done, the adverse event would not have happened) explains nothing about what actually happened or why.

In addition, counterfactuals are a powerful tributary to the hindsight bias. They help us impose structure and linearity on tangled prior histories. Counterfactuals can convert a mass of indeterminate actions and events, themselves overlapping and interacting, into a linear series of straightforward bifurcations. As the sequence of events rolls back in time, away from its outcome, the story of bad decisions and wrongdoing builds. What is noticed is that people chose the wrong prong at each fork, time and again—ferrying them along inevitably to the outcome that formed the starting point of the investigation (for without it, there would have been no investigation).

However, human work in complex, dynamic worlds is seldom about simple dichotomous choices (as in to err or not to err), and it certainly is never done backward in time. Bifurcations in actual healthcare practice are extremely rare—especially those that yield clear previews of the respective outcomes at each end. In reality, choice moments (such as there are) typically reveal multiple possible pathways that stretch out, like cracks in a window, into the ever-denser fog of futures not yet known. Their outcomes are indeterminate, hidden in what is still to come. In reality, actions need to be taken under uncertainty and under the pressure of limited time and other resources. What from the retrospective outside may look like a discrete, leisurely two-choice opportunity not to fail is from the inside really just one fragment caught up in a stream of surrounding actions and assessments. In fact, from the inside, it may not look like a choice at all. These are often choices only in hindsight. To the practitioners caught up in the sequence of events, there was likely no compelling reason to reassess a situation or decide against anything (or else they probably would have) at the point the investigator now finds significant or controversial. They were likely doing what they were doing because they thought they were right given their understanding of the situation and their pressures, goals, and ambiguities.

The challenge for an investigator becomes understanding how this may not have been a discrete event to the people whose actions are under investigation. The investigator needs to see how other people's decisions to continue on a course of action were likely nothing more than continuous behavior—reinforced by their current understanding of the situation, confirmed by the cues on which they focused, and reaffirmed by their expectations of how things would develop.

When counterfactuals are used in investigations, even as explanatory proxy, they often require explanations as well. After all, if an exit from the route to trouble stands out so clearly to us, how was it possible for other people to miss it? If there was an opportunity to recover, not to crash, then failing to grab it demands an explanation. Investigators often look for clarification in the set of rules, professional standards, and available data that surrounded people's operation at the time and how people did not see or meet that which they should have seen or met. Recognizing that there is a mismatch between what was done or seen and what should have been done or seen—as per those standards—it is easy to judge people for not doing what they should have done.

If practitioner behavior is contrasted with written guidance that can be found to have been applicable in hindsight, then actual performance is often found wanting; it does not live up to procedures or regulations. Investigations invest considerably in organizational archeology so that they can construct the regulatory or procedural framework within which the operations took place or should have taken place. Inconsistencies between existing procedures or regulations and actual behavior are easy to expose when organizational records are excavated after the fact and rules uncovered that would have fit this or that particular situation.

This is not, however, informative. There is virtually always a mismatch between actual behavior and written guidance that can be located in hindsight. Pointing out that there is a mismatch sheds little light on the why of the behavior in question. For that matter, mismatches between procedures and practice are not unique to mishaps. The behavior is then contrasted against the investigator's reality, not the reality surrounding the behavior in question at the time.

This also happens when investigators hold individual performance fragments in finding all the cues in a situation that were not picked up by the practitioner, but that in hindsight proved critical. This is a standard response after adverse events: Point to the data that would have revealed the true nature of the situation or patient's state. Knowledge of the "critical" data comes only with the omniscience of hindsight, but if data can be shown to have been physically available, it is assumed that they should have been picked up by the persons in the situation. The problem is that pointing out that it should have does not explain why it was not or why it was interpreted differently then. There is a dissociation between data availability and data observability (Woods, Dekker, Cook, Johannesen, & Sarter, 2010)—between what can be shown to have been physically available and what would have been observable given the multiple interleaving tasks, goals, attentional focus, interests, and background of the practitioner.

Judging people for what they did not do relative to some rule or standard does not explain why they did what they did. The investigator has gotten caught in what William James called "the psychologist's fallacy" a century ago (Woods et al., 2010): He or she has substituted personal reality for that of the object of study.

It appears that to explain failure, investigators often seek failure. To explain missed opportunities and bad choices, they seek flawed analyses, inaccurate perceptions, violated rules—even if these were not thought to be influential, obvious, or even flawed at the time. Remember how this search for people's failures is another well-documented effect of the hindsight bias: Knowledge of outcome fundamentally influences how we see a process. If we know the outcome was bad, we can no longer objectively look at the behavior leading up to it—it must also have been bad (Fischhoff, 1975).

RECONSTRUCTING THE HUMAN CONTRIBUTION TO AN ADVERSE EVENT

How can the perspective from inside the situation be captured so that an investigation generates meaningful results? The investigator is confronted with a problem similar to that of the field researcher—how to migrate from a context-specific set of data to more concept-based results that are interpretable and falsifiable, that are more than just another anecdote. Falsifiability means the investigator has to leave a trace that others can follow. In human factors, it is not uncommon to make the shift from a context-specific description to a concept-dependent one in one big leap, which produces conclusions that no one else can verify. For example, an investigation may conclude that after a poor clinical outcome, there was evidence of deficient medical decision making. But what exactly couples one to the other?

The challenge is to build an account that moves from the context specific to the concept dependent gradually, leaving a clear trace for others to follow, verify, and debate. To be sure, any explanation of past performance that the investigators arrive at remains a fictional story, an approximation, a tentative match—open to revision as new evidence may come in. In the words of Woods (1993):

> A critical factor is identifying and resolving all anomalies in a potential interpretation. We have more confidence in, or are more willing to pretend that the story may in fact have some relation to reality if all currently known data about the sequence of events and background are coherently accounted for by the reconstruction. (p. 238)

Human factors reconstructions typically follow a number of steps that the investigator could use to begin to reconstruct a concept-dependent account from context-specific incident data (Dekker, 2002; Xiao & Vicente, 2000).

The first step often consists of laying out the sequence of events in context-specific language. It is hoped that the record and other data about an incident reveals a sequence of activities—human observations, actions, assessments, decisions, as well as changes in the state of the process or system. This sequence of events forms the starting point for an examination of the inside of the situation. The goal is to examine how people's mindsets unfolded parallel with the situation evolving around them and how people, in turn, helped influence the course of events. Cues and indications from the world influence people's situation assessments, which in turn inform their actions, which in turn change the world and what it reveals about itself, and so forth. This means that if certain actions or assessments are difficult to interpret, then the circumstances (particularly what was observable about them) in which they appeared can hold the key to why they made sense. Indeed, the reconstruction of mindset often begins not with the mind but with the situation in which the mind found itself.

Similarly, if there is a lack of data from system or process sources, certain behaviors that are canonical in particular process states can help to reconstruct the state of cues and indications observable at the time. This creates various entries to scour the record for events and activities. Shifts in behavior represent one of those entries. There can be points at which people may have realized that the situation was different from what they believed it to be previously. This is seen in their remarks, notations, or actions. These shifts are markers to look at later for the indications unfolding around them that people may have used to come to a different realization. Actions to influence the process are another. These may come from people's own intentions. Depending on the kind of data that the domain records provide, evidence for these actions may not be found in the actions themselves but in process changes that follow from them. As a clue for a later step, such actions also form a nice little window on people's understanding of the situation at that time.

Changes in the process are yet another entry. Any significant change in the process that people manage must serve as an event. Not all changes in a clinical process managed by people actually come from the people managing it. Also, increasing automation in a variety of workplaces has led to the potential for autonomous process changes almost everywhere—for example, automatic shutdown sequences or other interventions, alarms that go off because a parameter crossed a threshold, uncommanded mode changes, and autonomous recovery from undesirable states or configurations. Yet even if they are autonomous, these process changes do not happen in a vacuum. They always point to human behavior around them: behavior that preceded the event and behavior that followed it. People may have helped to get the process into a configuration that triggered autonomous changes. When changes happen, people notice them or not; people respond to them or not. Such actions, or the lack of them, again give the investigator a strong clue about people's knowledge and current understanding.

The way to capture these events and activities during this stage is in context-specific language—meaning a minimum of psychological diction. Instead, use a version of what happened in terms that people in the domain use to talk about their work. The goal is to miss as few details as possible. Skipping to higher-level descriptions of human performance is seductive, even at this stage, but should be avoided. Seemingly low-level concepts, such as "decision making" or "diagnosis," already are large—meaning they contain a lot of behavior—and are easily mistaken for detailed insight into psychological issues.

Time (or space) can be powerful organizing principles to help lay out the activities and events. Behavior, and the process in which it took place, unfolded over time, and probably in some space. By organizing data spatially and temporally (e.g., through drawing maps, timelines, or both), actions and assessments can become more clearly coupled to the process state and location in which they took place; they can recover their spot in the flow of events of which they were part and that helped bring them forth. Such organization likely yields further clues about why actions and assessments made sense to practitioners at the time.

The second step is to divide the sequence of events into episodes, if necessary and possible. Adverse events do not just happen; they evolve over a period of time. Sometimes, this time may be long, so it may be fruitful to divide the sequence of events into separate episodes that each deserves further human performance analysis. Cues about where to chunk up the sequence of events can mostly come from the domain description derived, especially at discontinuities in human assessments, actions, or process states.

There is, of course, inherent difficulty in deciding what counts as the overall beginning of a sequence of events (especially the beginning—the end often speaks for itself). Since, philosophically, there is no such thing as a root cause, there is technically no such thing as the beginning of a mishap. Yet the investigation needs to start somewhere. Making clear where it starts and explaining this choice is a good step toward a structured, well-engineered human performance investigation.

One option often chosen is to take as the beginning of the first episode the assessment, decision, or action by people or the system close to the adverse event—the one that, according to the investigator, set the sequence of events in motion. This assessment or action can be seen as a trigger for the events that unfolded from that point.

Of course, the trigger itself has a reason, a background, that extends back beyond the adverse event—both in time and in place. The whole point of taking a proximal action or event as a starting point is not to ignore this background but to identify concrete points to begin the investigation into them.

The third step is to find out how the world looked or changed during each episode. This step is about reconstructing the unfolding world that practitioners inhabited, to find out what their process was doing and what data were available. This is the first step toward coupling behavior and situation—toward putting the observed behavior back into the situation that produced and accompanied it. Laying out how some of the critical parameters changed over time is nothing new to investigations. The record will most likely contain (some kind of) data about how process (or patient) parameters were changing over time and how these were presented to the people in question. Considerable domain knowledge (either from the investigator or from outside) may be necessary to determine which of the parameters could have counted as a stimulus for the behavior under investigation. The difficulty (reflected in the next step) will be to move from merely showing that certain data were physically available to arguing which of these data were actually observable and made a difference in people's assessments and actions—and why this made sense to them at that time.

The fourth step is to try to identify people's goals, focus of attention, and knowledge active at the time—the three sets of cognitive factors discussed in Chapter 3. So, of all the data available, what did people actually see, and how did they interpret it? Given that human behavior is goal directed and governed by knowledge activated in context, clues are available from looking at people's goals at the time and at the knowledge activated to help pursue them. Finding which goals people were working on does not need to be difficult. It often connects directly to how the process was unfolding around them. What was canonical or normal at this time in the operation? Tasks (and the goals they represent) relate in systematic ways to stages in a process. What was happening in the process managed by the people? What were other people in the operating environment doing? People who work together on common goals often divide the necessary tasks among them in predictable or complementary ways.

It is seldom the case, however, that just one goal governs what people do (see Chapter 3). Most complex work is characterized by multiple goals, all of which are active or must be pursued at the same time (e.g., on-time performance and safety). Depending on the circumstances, some of these goals may be at odds with one another, producing goal conflicts. Any analysis of human performance has to take the potential for goal conflicts into account. Goal trade-offs can be generated by the nature of the work itself. For example, anesthesiologists need to maximize preoperative workup time with a patient to guard patient safety and quell liability concerns, while their schedules interlock with other professions that exercise pressure with respect to, for example, timing. Goal conflicts can also precipitate from the organizational level. In this case, not all goals (or their respective priorities) are written down in guidance procedures or job descriptions. In fact, most are probably not. This makes it difficult to trace or prove their contribution to particular assessments or actions. However, previous occurrences in similar circumstances or in the same organization may yield powerful clues. They can substantially influence people's criterion setting with respect to a goal conflict.

When it comes to knowledge, not all knowledge people once showed to possess is necessarily available when called for (see Chapter 3). In fact, the problem of knowledge organization (is it structured so that it can be applied effectively in operational circumstances?) and inert knowledge (even if it is there, does it get activated in context?) should attune investigators to mismatches between how knowledge was acquired and how it is to be applied in practice. For example, if material is learned in neat chunks and static ways (books, most computer-based training) but needs to be applied in dynamic situations that call for novel and complex combinations, then inert knowledge is a risk. What people know and what they try to accomplish jointly determine where they will look, where they will direct their attention—and consequently which data will be observable to them.

Recognize how this is, again, the local rationality principle. People are not unlimited cognitive processors. People do not know and see everything all the time. So, their rationality is limited or bounded. What people do, where they focus, and how they interpret cues makes sense from their point of view, their knowledge, their objectives, and their limited resources (e.g., time, processing capacity, workload).

Reestablishing practitioners' local rationality will help in understanding the gap between data availability and what people actually saw or used. In dynamic situations, people direct their attention as a joint result of what their current understanding of the situation is, which in turn is determined partly by their knowledge and goals and what happens in the world. Current understanding helps people form expectations about what should happen next (either as a result of their own actions or as a result of changes in the world itself). Particularly salient or intrusive cues will draw attention even if they fall outside people's current interpretation of what is happening. Keeping up with a dynamic world, in which situations evolve and change, is a demanding part of much operational work and implies two different kinds of errors. People may fall behind during rapidly changing conditions and update their interpretation of what is happening constantly, trying to follow every little change in the world. Or, people may become locked in one interpretation, even while evidence around them suggests that the situation has changed (De Keyser & Woods, 1990).

Finally, the investigator needs to step up to a conceptual description of what happened in the adverse event. The goal here is to build an account of human performance that runs parallel to the one created in the first step. This time, however, the language that describes the same sequence of events is not one of domain terms; it is one of human factors or psychological concepts. The point of the final step is to close the gap from data to interpretation by following and documenting the various steps between the context-specific account of what happened and a concept-dependent one, linking back the concepts found to specific evidence in the context-specific record.

One reason for the importance of this step goes beyond the mandate of an individual investigation. Getting away from the context-specific details—which are set in a language that may not communicate well to accidents or problems in other domains—opens up a crucial way to learn from failure: discovering similarities between seemingly disparate events and even application areas. When people instead stress the differences between sequences of events or domains, learning anything of value beyond the one event becomes difficult. Similarities between accounts of different adverse events can point to common conditions that helped produce the problem under investigation. That, of course, is the whole point of having an adverse event investigation contribute to organizational learning: not only putting out the fires of individual events but also beginning to understand the underlying systemic factors that keep producing adverse events—and to find ways to change them.

Case Study 6.1: From Punitive Response to Organizational Learning

A few years ago, a colleague and I had the opportunity to find out more about the connection between confidential reporting and learning (Dekker & Hugh, 2010; Dekker & Laursen, 2007). Particularly, we wanted to examine whether it is chiefly the lack of retribution that makes people report more (and more useful information that promotes organizational learning) or whether there are additional intervening variables at work.

We were able to trace a safety-critical organization over a period of 2 years as it attempted to convert from line-management-driven punitive incident responses to a confidential reporting system run by the safety staff. The organization employed a total of 1,400 people, of whom 400 were frontline operators—those in direct operational contact with the safety-critical process. It had run up against the limits of the so-called blame cycle (Reason, 1997). Incidents had been seen as a result of human error, triggering reprimands and extra training for individuals, which often resulted in a repetition of the incident (but by a different operator) as basic working conditions were left unchanged.

While the organization thought it was doing what it could, the incident count did not go down. As it often does, this opened a window for new approaches, and the organization was interested in getting to know about different ways of dealing with incidents and their reporters as a possible route to greater learning. With guidance, a safety staff was set up and given a broad mandate for devising an incident report collection and analysis process. The basic transition was as follows (and happened about 6 months into the 2-year project):

Before, the employee involved in an incident had to report to his or her line manager, who would then devise corrective actions (mostly a reminder to watch out, some extra coaching, or retraining for the individual involved). Reporting was hardly voluntary; employees were compelled to report on their own safety performance problems because they knew that others who interacted with their safety-critical process would otherwise discover and report them—something that could lead to even harsher consequences.

After the transition, the employee could bypass line management and report the incident (on paper or in person) to a newly revamped safety staff (consisting of operators), who would then try to extract broader learning leverage from the reported occurrence, often together with the person involved. This person would not be connectable to the occurrence by anyone other than safety staff.

During the 2 years of this project, we interviewed numerous participants at different levels in the organization and were closely involved with the developing safety staff and its activities. Interviews were structured around 10 questions. Questions 1 to 7 were asked during interviews in the period before the transition. Questions 4 to 10 were asked after.

1. Describe the process of filling in the reporting form and what happens afterward. What feedback do you get? Did you get an interview with your line manager, and how did you experience that?
2. Describe the process of operational incident reporting and incident management within the company. How and when do you fill in a report?
3. Was there any focus on learning? If yes, how?
4. Accessibility of reporting forms?
5. Was it easy to fill in a report?
6. What did you see as the purpose of submitting a report?
7. Do you feel that your interview and the following report written about it really captured the essence of the incident?
8. Does the confidential process have an influence on your motivation to report incidents?
9. Have you observed any shift away from using reminders and procedures as countermeasures to achieve change within the organization?
10. Are compliance-based or behavior-directed programs still used as a means for making progress on safety?

Our main group of interview participants consisted of operators filing reports (both before and after the transition). We sought to answer how well operators liked the new reporting scheme; what they learned from participating in it (if anything) that they did not know before; what changes in people's job behavior occurred that could be linked to the new reporting scheme; and whether there were any other tangible results from it, particularly in terms of producing greater leverage for organizational learning. For the last purpose, we also reviewed considerable archival material, particularly incident reports written inside the organization, to learn more about the conceptualization of risk sources and proposed countermeasures before and after the transition.

The findings, unsurprisingly, confirmed that fear of retribution hampers safety reporting. When the organization shifted from line-management-based evaluations of reports to a confidential safety staff dealing with reports, the number of reports increased. People's reported willingness to send them in also increased, as did the relevance and resolution of their content.

But more seemed at play. Before the transition, employees actually were ready to confess an "error" or "violation" to their line manager. It was almost seen as an act of honor. Reporting it to a line organization—which would see this as a satisfactory conclusion to its incident investigation—produced rapid closure for all involved. Management would not have to probe deeper; the operator had seen the error of his or her ways and had been reprimanded and told or trained to watch out better next time. For the operator, simply and quickly admitting an error avoided even more or deeper questions from the line manager and could help avert career consequences, in part by avoiding passing information up or on to other agencies (e.g., the regulator of the industry).

Fear of retribution, in other words, did not necessarily discourage reporting. In fact, it encouraged a particular kind of reporting: a mea culpa with minimal disclosure that would get it over with quickly for everybody. Both reporters and management engaged smoothly in what Erik Hollnagel called an efficiency-thoroughness trade-off (ETTO), in this case all efficiency, no thoroughness (Hollnagel, 2009). Get the aftermath of an adverse event over with, achieve closure. "Human error" as cause seemed to benefit everyone—except organizational learning.

Here is an example of what one practitioner told us:

> I didn't tell the truth about what took place, and this was encouraged by the line manager. He had made an assumption that the incident was due to one factor, which was not the case. This helped me construct and maintain a version of the story, which was more favorable for us practitioners.

In the few cases for which reports of errors did go up the line into the organization before the transition, directives typically came back exhorting practitioners to watch out more carefully for that particular problem or to adhere more stringently to a rule or guideline that already existed. What lacked was the notion that organizational learning through reporting happens by identifying systemic vulnerabilities to which all operators could be exposed, not by telling everybody to pay more attention because somebody did, on one occasion, not do so. Only by constantly seeking out its vulnerabilities can an organization develop and test more robust practices to enhance safety. But this puts a particular premium on what kind of incident reports—and what kind of reporter treatment—would be useful.

If learning hinges on the ability to dig out systemic vulnerabilities, then reports and organizational encounters with reporters need to go beyond the phenotypical "errors" or "violations" that may have served as the report's trigger. They need instead to engage the so-called second stories (Woods et al., 2010). The distinction between first and second stories of failure has been useful in driving change across several domains, including healthcare (AMA, 1998), and it also provided a good hinge in ours. First stories reveal how an outcome could simply have been avoided if the people involved had invested a little more effort or had been more careful. They fall back on "human error" as explanations and stop there, making people and organizations wonder how they can possibly cope with the unreliability of the human element in their midst.

Here is an example of a first story—a deidentified organizational memo documenting the countermeasures after a particular incident:

> The incident has been discussed with the concerned practitioner, pointing out that priorities have to be set according to their urgency. The operator should not be distracted by one single problem and neglecting the rest of his working environment. He has been reminded of applicable rules and allowable exceptions to them. The investigation report has been made available to other operators by posting it on the intranet.

Here is another:

> Head of operations interviewed the practitioners after the incident. They were reminded about correct and safe planning as well as good monitoring of their process in case of a slightly tight situation.

In first stories, personal attributions are typically made to help explain why things went wrong (e.g., a line manager blaming an operator's "aggressive attitude"). Second stories, in contrast, make different attributions to find out why things go wrong. They reveal the multiple conflicting goals, pressures, and systemic vulnerabilities beneath the "error" to which everybody in the system is exposed. Second stories use human error as a starting point, not as a conclusion. Digging for second stories is crucial to learning as it promotes the discovery of systemic vulnerabilities. Recognizing these is a precondition for making organizational investments to cope with the real sources of risk: the genotypical contributors to failure.

In some cases before the transition, safety improvements were thought to result from getting rid of "bad apples" who contaminated or undermined an otherwise safe system. Individuals were seen as sole sources of failures and problems. As per one memo:

> The involved trainee has been terminated, he is not working as a practitioner any more. His incident will cause further investigation about roles and responsibilities and may lead to disciplinary sanctions.

After the transition in the organization we studied, such individually oriented countermeasures became rare. Incident reports and investigations came up with deeper contributory sets that could not be ignored and that took line management into different areas than before the transition. Learning became possible because systemic vulnerabilities had been identified, reported, studied, contextualized, and checked against relevant practitioner expertise.

After the conversion to a confidential system run by the safety staff, the safety investigation reports written on the basis of operator interviews and other data typically began to contain a larger set of contributory factors. They also shed language such as "the operator should have ... " or "if only the operator had ... ," instead trying to probe the reasons why it made sense for operators to do what they did. This would automatically offer an entry door into second stories, as investigators were forced to dig deeper into the organization for systemic reasons behind practitioners' performance. Simple causal statements gradually made way for more complex etiologies that could take an entire paragraph. Operators felt that levers for organizational learning were being identified, in sharp contradistinction with the previous regime. Here is an example of a spontaneous reaction: "I congratulate you with this report. I only hope that your suggestions will be heard and actions will be taken at higher echelons. This way we can all profit from one incident."

Getting to second stories is clearly a precondition for finding these leverage points and making systemic changes to working circumstances. But this requires that incident reporters are met not only in a nonjeopardizing setting, but also by somebody who understands his or her work, who can ask the right questions and ask them legitimately and enter into a meaningful dialogue to jointly discover more. Of course, identifying systemic leverage points does not guarantee organizational learning, but it represents a precondition for learning.

The shifting nature of interviews with the recipients of incident reports (first, line managers who may have entertained a distant view of real practice; then safety staff, consisting of operators closely connected to the actual nature of work) introduced an element considered key in the creation of a safety culture: employee empowerment. Offering practitioners the opportunity to contribute actively to the conceptualization of risk and the search for systemic vulnerabilities underlying it appears to be motivating. In fact, interviews revealed that the chief reason why practitioners' willingness to report increased was not the lack of retribution but rather the realization that they could make a difference.

Giving operators the leverage and initiative to help achieve safety gains turned out to be a large motivator to report. It gave them part ownership in the organization's safety record. An important factor for this to work did turn out to be the legitimization of questions about operator performance and the context in which it occurs. In the organization studied here, that was done by having the safety staff consist (in part) of practitioner employees:

> It is very good that a colleague, who understands the job, performs the interviews. They asked me very good questions and pointed in directions that I hadn't noticed. It was very positive compared to before. Earlier you never had the chance to understand what went wrong. You only got a conclusion to the incident. Now it is very good that the report is not published before we have had the chance to give our feedback. You are very involved in the process now and you have time to go through the occurrence. Before you were placed in the hot seat and you felt guilty. Now, during interviews with the safety staff, I never had the feeling that I was accused of anything.

Before the transition, organizational learning was thought to be accomplished through reminders and reprimands and through the top-down dispensing of awareness about a problem to which a particular operator had been exposed. While raising awareness of safety problems is not thought to have any sustained effect, results here indicate that it can have such an effect, but only under near-perfect circumstances. Particularly, awareness should be raised by a peer, somebody who has legitimacy and knowledge to speak about the issue. It should be specific enough to target recognizable situations. Discussions work much better than posters. One-on-one instruction works even better. A sustained effect also demands follow-up and appropriate repetition.

While the results from this case study do not contradict the basic wisdom of confidential reporting, they suggest that employees' willingness to report hinges on more than a lack of fear of retribution. The results identified a more complex relationship between retributive probability and reporting. In the old, punitive system studied here, employees were actually eager to report (a particular version) precisely so that they could get off the hook. Our results showed that willingness to report could be mediated less by a fear of retribution and more by a feeling of empowerment, of being able to cooperate in creating organizational safety, to feel ownership, a stake, or coresponsibility for the safety record. The transition reported here gave employees precisely that, something that not only triggered congratulatory comments from operators but also actually provided the organization with new leverage points for learning.

HUMAN FACTORS AND RESOURCE MANAGEMENT TRAINING

The realization that most adverse events are triggered not by a lack of technical prowess or knowledge but by the breakdown of coordination between people (and people and technology) has been a watershed in safety thinking. It has increasingly focused the attention of educators and safety professionals on the soft skills that hold hard, technical work together.

For some 30 years, disasters in sociotechnical systems have been understood to be the result of a concatenation of multiple, small, and separate events that become linked and amplified. Eventually, they push the system over the edge into disaster (Reason, 1990; Weick, 1990). When taken separately, these small events are incapable of creating much havoc, and even in combination they do not always have to produce disaster. In fact, many of these small events and problems happen in safety-critical worlds every day. They are part of normal practice. The point of incident reporting and adverse event investigation, discussed in this chapter, is to understand how these small events combine to create greater trouble.

What links these small events more than anything else, is human decision making and coordination. That is what, in complex and unforeseeable ways, can string events together in a trajectory toward bad outcomes. To be sure, human decision making and coordination are also necessary *not* to have the small events combine into larger problems, to prevent and stop progressions into worse situations. But the kind and quality of such coordination and decision making is critical—often considerably more critical than the mere possession of skills or knowledge by some human member of the problem-solving team.

Retrospective analyses have often been able to show that the knowledge or understanding of a problem was present somewhere in, or scattered across, the human and machine problem solvers who were present at the time. But breakdowns in coordination among them made it so that the solution never became apparent. These breakdowns themselves can take various forms, of course, and include

> the interruption of important routines, regression to more habituated ways of responding, the breakdown of coordinated action, and misunderstandings in speech-exchange. … When these four processes occur in the context of a system that is getting more tightly coupled and less linear, they produce more errors, reduce the means to detect those errors, create dependencies among the errors, and amplify the effect of those errors. (Weick, 1990, p. 572)

Such processes have been referred to (among others) as the "human factors" that undermine well-engineered safety-critical systems. They interact with the tightening of coupling in systems and their growing complexity (see Chapter 5). Tight coupling and interactive complexity make understanding and communicating more difficult, while miscoordination can also increase coupling and complexity and so on. Weick (1990) explained, for example, how the consciousness of a group can become an arena for troubleshooting that is devoted to a higher-order question (e.g., the next patient on the surgical list, a bed shortage) rather than the operation at hand. The higher-order plan interrupts and uses up attention and other cognitive resources that could have been allotted to the immediate, lower-order issues to be solved or addressed (Weick, 1990).

This, however, is only half the story. These same processes are also invariably responsible for the creation of success and safety in changing, complex settings. The ability to divert attention to new cues and not get fixated on one problem is the other side of interruptions. By habitually (and thus efficiently) responding to more familiar aspects of a problem, cognitive resources (for thinking, talking, attending, decision making) can be saved for more difficult aspects of the problem. Communication and coordinating action can prevent tight coupling and complexity from going out of control.

This insight has been fundamental to the growth of so-called crew resource management programs, first in aviation and then in a range of other industries, including shipping, nuclear power, and lately healthcare. Effective human coordination (and human-machine coordination) is often, if not always, the ingredient that can make for successful outcomes even in tightly coupled and complex situations. But that does put a particular premium on the kind of human coordination (and, again, human-machine coordination).

COMMUNICATION AND COORDINATION BREAKDOWNS

So, how does human communication become ineffective? How can coordination break down, and for which signs should we be looking for? A research direction that has shown promise has been developed by social scientists such as Nevile (Nevile, 2004; Nevile & Walker, 2005) using microanalytic techniques like conversation analysis to study naturally occurring interaction. There are few settings in which communication in healthcare is recorded (unlike in aviation). But the findings by Nevile and others point to the places and practices where coordination may be breaking down or may already have broken down. Having some idea of this may sensitize other practitioners to some of the vulnerabilities their own coordination may be facing.

In a conversation analysis of the cockpit voice recording of a particular accident flight in Australia, Nevile and Walker (2005) showed how coordination between the two crew members broke down before the aircraft crashed on an approach to a runway in bad weather. The analysis offered in their special report is canonical in the sense that it is one of the first (but see also Predmore, 1991) that imported social scientific analysis into the microdetails of a voice record in an attempt to substantiate any claims about the quality of the coordination between practitioners that occurred. Their analysis identified at least three important aspects of interaction between practitioners.

First, in the Australian accident analyzed by Nevile and Walker (2005), there were many instances of overlapping talk. That is, both practitioners would speak at the same time. This obviously is problematic for coordination. It forces participants to merge the roles of speaker and listener, and things easily get lost as a result. Overlapping talk occurs when somebody other than the original speaker starts to talk. Another study found that this is actually unusual for practitioners at work; they normally wait with speaking before the other speaker is entirely finished. Overlapping talk does occur occasionally, particularly in highly pressurized and high-tempo activities. It may also occur occasionally in dialogue that is not task oriented. If overlapping talk happens routinely in interaction with another practitioner, however, it could signal trouble and may be a reason to call for a brief halt in operations to reflect and restart.

Second, in this Australian accident, there were many instances when the captain (or senior crew member) of the aircraft said something, and the first officer said nothing in reply, even though some kind of response would have been appropriate, relevant, and an expected next action. Overwhelmingly, interactive conversation normally involves the production of talk in response to other talk. One person says something; the other says something that may be deemed appropriate as a follow-up or response. This conversation is formally called the first-pair part and the second-pair part of conversation, respectively. People are sensitive to the sequential nature of conversation. Silences, even short ones, are meaningful and might signal a problem with the first-pair part (e.g., it was not heard or was deemed irrelevant or inappropriate) (Nevile, 2002).

Third, the analysis of the interaction showed how the captain often corrected, or repaired, the first officer's remarks, even when there was no sign of any problems, from the first officer's point of view, in his actions or utterances. This is an important sign of coordination breaking down. *Repair* refers to those points in conversation when participants try to recover from communicative problems of some sort. There is a marked tendency in everyday conversation for self-repair (Nevile, 2004). Even when the other person initiates the repair (which may be done with certain body language), the speaker will mostly repair his or her talk as well. In the more exceptional cases when repair is both initiated and executed by another person in the conversation (itself unusual, it is called other initiated other repair), then this is often done in such a way (hedging, qualifying, delaying, softening) to smooth the impact of the repair for the recipient.

So, how can coordination, particularly across hierarchies (between junior and senior practitioners) avoid producing dysfunctional patterns? This has been the question taken on by crew resource management training in aviation and beyond. Its initial mandate was to teach and help junior members of a team to speak up, and there was good reason to try to do something about that.

Speak Up

In June 1972, flight BAE 548, a Trident airliner scheduled to fly from London Heathrow to Brussels, crashed at Staines just south of the London airport not long after takeoff. The crew consisted of Captain Stanley Key, First Officer Jeremy Keighley, and Flight Engineer Simon Ticehurst. The captain was 51 years old and had 15,000 hours of flying experience, of which 4,000 were on Tridents. Keighley was only 22 and had joined line flying a month and a half earlier, with 29 hours on the aircraft. Ticehurst was 24 years of age and had over 1,400 hours, including 750 hours on Tridents.

The doors closed at 1558, and at 1600 Key requested to be pushed back from the gate. At 1603, BE 548 was cleared to taxi to the holding point adjacent to the start of the runway. During taxi, at 1606, the flight received its departure route clearance, a routing known as the Dover One Standard Instrument Departure.

The aircraft left the ground at an airspeed of 145 knots (269 kilometers/hour). It reached a safe climb speed quickly, and the undercarriage was retracted. After 19 seconds in the air, the autopilot was engaged at 355 feet (108 meters) and 170 knots (310 kilometers/hour); the autopilot's so-called airspeed lock was engaged even though the actual required initial climb speed was 177 knots (328 kilometers/hour). Passing 690 feet (210 meters), Key commenced the turn toward the Epsom beacon and reported that he was climbing as cleared, and the flight entered cloud.

Then, Key commenced a standard noise abatement procedure that involved reducing engine power. As part of this, he retracted the flaps from their takeoff setting of 20°. Shortly afterward, BE 548 reported passing 1,500 feet (460 meters) and was recleared to climb to 6,000 feet (1,800 meters). During the turn, the airspeed decreased to 157 knots, 20 knots below the target speed. Then, the leading edge flaps were retracted as well. While this is necessary to be able to increase the speed of the aircraft, retracting flaps makes it harder for the wing to carry the aircraft at lower speeds.

Multiple warnings sounded that the aircraft was entering a stall (a situation when the angle of the wing relative to the airflow is such that the wing stops producing lift, and the aircraft no longer flies; this often happens when airspeed is too low), and an automatic recovery system shook the control column in Key's hands and then tried to push it forward. But Key continued to hold back, trying to slow the aircraft and aggravating the stall.

Eventually, the aircraft entered what is known as a deep stall—a situation in which the horizontal tail plane is enveloped in the disturbed airflow from the stalled wings (a design such as the Trident is particularly vulnerable to this due to the placement of the horizontal tail plane). In a deep stall, the tail plane can no longer help the airplane regain airspeed (by pushing the nose down) as it itself has been rendered useless by the disturbed airflow. The airplane crashed near the town of Staines.

The flight had lasted just over 2 minutes. Everybody on board was killed (118 people). None of the other crew members intervened with the apparent actions by Key that endangered the flight. The reason for why Key was doing things the way he was became clear after the postmortem. A pathologist concluded that Captain Key had atherosclerosis and had suffered a potentially distressing arterial event, or heart attack, caused by raised blood pressure typical of stress. Not long before the flight, Key had been embroiled in a heated argument in the crew room with a colleague pilot about a disputed strike (which he was vehemently against). It is possible that Key suffered the heart attack at around the time of commencing noise abatement procedures. Although the pathologist could not specify the degree of discomfort or incapacitation that Key might have felt, it may well have interfered with his ability to make decisions, to see things, or even to remain conscious.

What remained a greater mystery was why none of the other crew members apparently intervened in Key's decision making and actions. Many copilots were intimidated by his demeanor, his experience, and his status. A flight engineer had even carved graffiti in his control station behind the pilots: "Key has got to go." But when Key indeed went as a result of his heart attack, nobody in the same cockpit dared to intervene to save the flight.

High-visibility accidents such as that with BAE 548 have moved the problem of juniors not speaking up to seniors to center stage. Accidents like this reveal that the knowledge to solve a problem or reduce a risk was present among one or more members of the team, but that this knowledge (or its relevance or importance) was never successfully brought to the attention of the superior. It would seem that junior staff would rather pay with their own lives than upset their boss. And in healthcare, they may pay with the patient's life.

The reasons for this have been the subject of much study since the 1970s. In the United States, NASA was instrumental in providing research funding and creating the conditions for simulator experiments to tease out the breakdowns in interaction that led to trouble or even accidents (Billings, 1996; Fischer, Orasanu, & Montvalo, 1993; Orasanu, 1990; Orasanu & Martin, 1998). Other research centers, airlines, and universities around the world have led similar efforts or contributed to the increasing understanding of how hierarchical work relationships interfere with effective communication and problem solving (International Civil Aviation Organization, 1998; Orasanu, 1990; Predmore, 1991; Salas, Wilson, & Burke, 2006).

Orasanu and Fischer were able to categorize the data of the simulator experiments according to the level of mitigation of speech (Fischer et al., 1993). *Mitigation* means reducing the severity, seriousness, or painfulness of something. Copilots did most of the mitigation, and by doing so, they lessened the impact of their observation or opinion, sometimes to the point of going unheard altogether. In the most extreme cases of mitigation, a flight crew could effectively be reduced to a single pilot (which for those transport category airplanes is technically illegal—they may only be flown by a crew of two or more except in emergencies).

The six levels of mitigation are as follows (with healthcare examples for clarification). You can make or give a

1. *Command*: "Give him 5 milliliters now!" A command basically does not mitigate. There is no ambiguity regarding what one person wants the other to do and no dressing it up in any social nicety. It is extremely unlikely that a junior practitioner will address a senior in this way, except when there is great familiarity with each other and each other's work, for instance.

2. *Joint obligation statement*: "I think he needs an extra 5 milliliters." This is a statement, an observation, not immediately a call for somebody to act. The "I think" part is also a mitigation, of course—rather than demanding or firmly believing or seeing, you only think.

3. *Joint suggestion*: "Why don't we put in an extra 5 milliliters?" Note the use of "we." This suggests that you are in this together, that there is some joint responsibility for problem formulation, action, and outcome. It is no longer clear, however, whom you think really needs to take the action. That may easily leave things undone.

4. *Query*: "Would you like to put in some extra now?" With this mitigation, the speaker is conceding quite explicitly that he or she is not in charge. It might even be taken as an invitation to receive some instruction that will set the junior straight instead of an invitation to be an equal part of the problem-solving team.

5. *Preference*: "I think it would be smart to consider an extra 50 milliliters." This is not a question or observation, and it is certainly no call to action. In fact, it is a call for reflection, or consideration, and only expresses a preference for what should be considered. This mitigates speech even further, and the urgency has been taken out of it altogether.

6. *Hint*: "Doesn't he seem a little paler than usual?" The specificity is now entirely lost, as is any sense of urgency or suggestion of what action should be taken. A hint is by far the easiest to ignore entirely by the receiver. In Orasanu's and Fischer's (Fischer 1993) work, junior practitioners gave a lot of hints and seldom migrated up the mitigation steps to sharpen the urgency or specificity of their attempted interventions. The fatal accidents Orasanu and Fischer studied were frequently preceded by such hints from a junior to a senior practitioner.

These levels of mitigation do a successively worse job of persuading the other practitioner that you have something of value to say or add, which is an interesting (and in a sense, dehumanizing) feature of the work environments that we have constructed for people. Rules of politeness, address, and implicit expectations of showing and receiving respect can fit badly with the time dependence, data richness, and safety criticality of much healthcare work. Many of the behaviors that we as humans have collectively evolved to lubricate social interaction, and ensure our acceptance and perhaps even survival in social structures, in other words, are entirely inappropriate in a context of data overload and decision ambiguity, of rapidly developing anomalies and uncertain outcomes.

The setting of much healthcare work, then, requires that we unlearn that which makes perfect sense in most other social contexts and has probably helped us gain favorable reception and the reward of belonging to social groups in the rest of our lives. Such unlearning is extremely difficult. But think of it as a role in context. It is not that you *are* impolite, brusque, disrespectful, or irreverent. It is the context—a dynamic world full of cues and indications and risky traps—that calls for a role in which you *perform* that way not for the sake of your own survival (though in Orasanu's cockpit work (Fischer 1993) and indeed all flying, this definitely is a factor), but for somebody else's. As soon as you are out of this context, you do not have to behave like that anymore.

So, how can you avoid mitigating your speech? NASA and its contractors, in a completely different context than the cockpit research (in this case, it was space shuttle building and launching), had developed guidance for how to make an assertive remark if an employee noticed evidence of an anomaly or problem that might have an impact on shuttle safety. The implementation of the guidance was spotty and difficult, and its use might have eroded over time as relentless cost-saving and production pressures increased. But its steps are as relevant as ever, also in healthcare situations when a junior needs to get heard by a senior (Columbia Accident Investigation Board, 2003):

Opening: Get the person's attention. The volume and tone is important here, although it may depend on the perceived criticality of the situation. How to address superiors is often a matter of professional and cultural convention or determined in part by the immediate problem-solving setting (first name, title, last name, job role). Some hospitals have basically ruled out the use of titles and insist on first names only. The kind of appellation, however, may matter less than getting the person's attention unambiguously.

Concern: State the level of concern. You might simply say, "I have a concern" or "I have an immediate concern" or a "grave concern" for that matter. Never qualify the statement down. If you call it a "slight concern," then the point of calling attention to it may not be so obvious anymore, and you have already lost ground. Any qualification down (making the concern sound less important) will get the other person to believe that there is more margin to ignore the concern than you might want there to be. You could, however, think about calling it a concern from *your* point of view—a fair qualification that may invite more consideration from the other person. The risk, of course, is that your point of view is dismissed as underqualified or uninformed, and that only a junior person could see this as a concern. But if you are clear with the next steps, you may be able to deal with that.

Problem: State what you believe the problem is. Make clear what you believe is being overlooked or not done or done in contravention of some guideline or procedure you consider to be applicable at that moment. You may simply be noticing some kind of symptom that you suspect is not seen by the other person. Of course, you might be unaware of good reasons for the other person not to be paying attention to the symptom. Thus appearing stupid is the risk of appearing attentive. If this is the case, you can always convert the problem statement into an invitation to the other person to teach you something new. Saving your face, however, can never be as important as saving the patient.

Solution: If you have a solution in mind, suggest it. Be concise because the person you are talking to is likely engaged in various other tasks or has put them on the back burner only momentarily. The solution might be as simple as taking an extra look at something or as radical as breaking off the entire operation. Coming up with at least some kind of solution shows that you not only mean business with your concern, but also that you are committed to being a meaningful part of the team.

Agreement: Assertively but respectfully ask for a response from the other person if you have not gotten one so far. This may be as simple as asking "What do you think?" or "Don't you agree?" Getting agreement (or getting confirmed that you do not) forms an important component of all coordination in complex, dynamic settings: You want communication to be closed loop. This means that an acknowledgment or some other kind of response should follow your own speech. You can ask for it if it does not. Otherwise, you may remain in the dark about whether things were understood or will be done at all.

If all this fails, various other avenues may still be open. One is to appeal to rules or guidelines that might be appropriate for the situation. Another can be to see if you can get allies among other team members. The social cost of your intervention may go up as a result of this, of course. Appealing to guidelines or enlisting more team members may not endear you to the senior person any more than you have already done.

What matters, however, is not how you are liked (even if this has potential career consequences). What matters is your duty ethic to the patient who is at stake. If you do not speak up for the patient, who will? Not speaking up when you believe you should will carry its own career consequences. Perhaps it may not be as noticeable as a missed promotion or graduation, but it could well be a burden of shame that you will bear for the rest of your (professional) life.

THE FALLACY OF SOCIAL REDUNDANCY

One of the paradoxical findings of the NASA work was that accidents are more likely when the captain does the flying (Orasanu & Martin, 1998). Recall from the accident of BAE 548 that it was indeed Captain Key who was doing the flying. Pilots typically split the flying duties 50/50. On one leg, one of them will be the so-called flying pilot and the other the nonflying pilot. These duties will then be reversed on the next leg. The nonflying pilot has all kinds of duties, such as controlling the radios and a host of other switches, knobs, and handles for which the pilot who is flying has no time or spare hands. But a crucial part of nonflying duties is to monitor the flying pilot carefully for what he or she is doing and to correct and intervene when necessary.

Accidents are far more likely when the captain is the flying pilot. The mitigation of speech, which copilots engage in much more than captains, can explain this paradox. The ability to point out problems or intervene is restricted when the junior practitioner needs to act toward the senior one—much more so than when the roles are reversed. Having somebody with the most experience directly at the controls, then, may be more dangerous than having senior practitioners overseeing a junior who is executing the task (Weick, 1990).

This suggests that there is a fallacy of social redundancy (Sagan, 1993). In engineering, unreliable parts can be compensated for by putting in multiple unreliable parts. The probability that both will fail at the same time is much smaller than that one might fail (but then there is still the other). Some of this logic carries over to people, of course. But even a complete failure of Captain Key (because of his heart attack) did not allow the other (redundant) parts to kick into action and take over control. There was too much of an accumulated social barrier for that. The redundancy metaphor only goes so far for social systems, then.

Transplanting ideas that work well in engineered systems (redundant parts that create duplication or at least some overlap of functions) to social settings may create unexpected side effects. In social systems, the very presence of other parts may actually influence the reliability of each part. That is not the case for inert, inanimate parts in an engineered system. But the very presence of Captain Key (whether he was still functioning or not) ground down the reliability of the remaining components (pilots) in the cockpit.

Other cross-influencing effects of social redundancy are also possible. With multiple people looking at an unfolding situation, individuals can become paralyzed to act. They may assume that there are enough others who might act first if necessary, with the effect that nobody does anything. Research has also shown the tendency of groups of decision makers to gravitate toward more extreme solutions than individuals might and stick with them despite evidence of their undesirability. Having been dubbed "groupthink," the psychological drive for consensus at any cost in a cohesive decision-making team can suppress disagreement and prevent the generation and appraisal of alternatives (Janis, 1982). Social redundancy, in those cases, is a fallacy.

And of course, social redundancy can create much social interaction (even if this is entirely task oriented). This itself can interfere with task completion. As Weick (1988) observed after the Three Mile Island (TMI) nuclear accident in the United States:

> The recommendation that people should keep talking is not as simple as it appears, because one of the problems at Three Mile Island was too many people in the control room talking at one time with different hunches as to what was going on. (p. 589)

Having multiple people involved in problem solving automatically creates a coordination overhead. Cognitive resources are spent on exchanging information among team members; otherwise, the resources be available for other activities. This confirms the importance of coordinating work beforehand, so that in more tightly coupled and time-pressurized situations people can rely at least to some extent on predetermined roles and heuristics. Fighter pilots often do this in pre-flight briefings, so as to avoid having to engage in deliberative decision making on the fly—when there is no time or cognitive space to do so (Svenmarck & Dekker, 2003). Weick (1988) offered the same reflection on the problem-solving work involved in TMI:

> The din created by tense voices plus multiple alarms, however, would make it all but impossible to single out talk as uniquely responsible for confusion, misdiagnosis, and delayed responding. The crucial talk at TMI should have occurred in hours before the control room got cluttered, not after. (p. 589)

Preoperative briefings (even for procedures or settings that have nothing to do with surgery) can be an enormously effective investment in safety precisely because they get some of the hard coordination and decision making over with before the work really picks up speed. They also help remind people of where to allocate attention during specific times of a procedure or shift ("watch this carefully when ... ") and can clarify roles ahead of time. The preoperative briefing can, or should, be the time when senior practitioners make their expectations of junior scrutiny and intervention explicit. This helps create the social and professional space for juniors to speak up by the time all team members are deeply involved in their shift or procedure. Such briefings are increasingly a part of the protocol or time-out before, for example, surgery (see "Briefings and Checklists" in this chapter).

DIVERSITY

Research on complex systems showed that collections of individuals with diverse tools and backgrounds can outperform collections of high-ability individuals (Page, 2007). Why is diversity so important? Effective management of difficult situations requires the ability to recognize, adapt to, and absorb problems and disturbances without noticeable or consequential decrements in performance. Diversity is a critical ingredient for this because it gives a system the requisite variety that allows it to respond to disturbances. With diversity, a system has more perspectives to view a problem and a larger repertoire of possible responses. Diversity means that routine scripts and learned responses do not get overrehearsed and overapplied, but that an organization has different ways of dealing with situations and has a rich store of perspectives and narratives to interpret those situations.

High-reliability organizations have been presented as an example of this (see Chapter 5). Such organizations show considerable decentralization of decision-making authority about safety issues. This permits rapid and appropriate responses to dangers by the people closest to the problems at hand. Wildavksy (1988) called this "decentralized anticipation," which emphasizes the superiority of entrepreneurial efforts to improve safety over centralized and restrictive top-down policies or structures. High-reliability theory has shown the need for and usefulness of operational discretion at lower levels in the organization. It has found collegial processes at work, with considerable operational authority resting at low levels in the organization in applications ranging from nuclear power plants to naval aircraft carriers and air traffic control. Even the lowest-ranking individual on the deck of an aircraft carrier has the authority (and indeed the duty) to suspend immediately any takeoff or landing that might pose unnecessary risk (Rochlin, LaPorte, & Roberts, 1987).

As a result, recommendations made by high-reliability theory include a validation of minority opinion as well as an encouragement of dissent. This can translate into the organizational empowerment of lower-ranking employees, such as nurses in hospitals (Pronovost & Vohr, 2010). The problem of not doing so, said Pronovost and Vohr, "starts in medical school and nursing school, where doctors aren't trained to listen to nurses and nurses aren't trained to ask questions or believe that they play an important role in the patient's care" (p. 109).

With stronger contributions from below comes more diversity, and more diverse teams are generally better prepared to deal with unexpected developments and situations (Starbuck & Farjoun, 2005). It creates what is known as superadditivity, or a total that is more than the sum of its parts, particularly when interactions between ideas is encouraged. People who think alike get stuck (Page, 2007). Diversity, then, can be seen as a way to circumvent the fallacy of social redundancy. The greater the difference between the parts is, the greater the likelihood of superadditivity there will be and the less likely it is for a team to become fixated on a single solution that may not be the right one.

Intervention decisions in safety-critical monitored processes are notoriously difficult. These decisions, particularly in escalating circumstances, are extremely hard to optimize, even with experience accrued over time. If the human operator waits too long with diagnosing a problem and gathering evidence that supports an intervention, the decision to intervene may well come too late relative to problem escalation. Degradation has then taken the process beyond any meaningful ability to reestablish process integrity. The human operator can draw few conclusions other than that intervention was too late (but it is seldom clear how much too late). If, in contrast, the human operator intervenes too early, then any evidence that a problem was developing that warranted intervention may disappear as a result of the intervention. The human operator, in other words, is left with no basis to learn about the timing of his or her intervention. How much too early was it? Was the intervention really necessary?

One area in which this dilemma of intervention presents itself acutely is in obstetrics. The intrapartum is surprisingly dangerous. For a variety of anatomical and physiological reasons, even normal labor is hazardous to both parturient and fetus. Uterine contractions impair placental oxygenation as a normal aspect of the intrapartum, something for which the fetus can compensate to some extent through peripheral vasoconstriction, anaerobic metabolism, and glycogenolysis, the process of internally generating energy for particularly the brain to survive in conditions of hypoxia.

The problems of supervisory control and intervention in escalating situations (e.g., growing fetal distress) present themselves acutely in obstetrics: Interventions of various kinds are possible and called for on the basis of different clinical indications and can involve everything up to a decision to deliver the baby by emergency cesarean section. The difficulty in timing these interventions right is apparent in the increase of even nonemergency cesarean sections in various countries despite the lack of persuasive clinical indications regarding their necessity and postoperative risks to the mother. The growth of a malpractice insurance crisis in a state like New Jersey (where insurance quotes for obstetricians have in some cases gone up to US$200,000 per year) has pushed many practitioners of obstetrics-gynecology to limit their work to gynecology (Zaccaria, 2002). The acrimony in New Jersey in this regard attests to the difficult and contested nature of obstetric decisions on how and when to intervene in pregnancy and labor and how these decisions are constructed in hindsight.

Adding team members to an obstetric escalating situation does not necessarily increase diversity because of the differential weighting of the voices that get added. We would need some type of diversity index to explore the resilience of various obstetric team makeups. One candidate type of diversity measure is the Herfindahl index of market concentration, which tries to map how the market shares of different organizations in a particular branch provide competitive diversity or rather monopolistic power. A high index is close to a monopoly; a low index means great diversity. The Herfindahl index H is computed as follows:

$$H = \sum_{i=1}^{N} s_i^2 \qquad (6.1)$$

where s is the decision share of each participant or the percentage of the voice they have in making an intervention decision.

If we apply this to an escalating situation in obstetrics, we can imagine the following: Suppose that the situation begins with two midwives, who both have a 45% share in any intervention decision that is made. There is also a lower-ranking nurse present, who has, say, a 10% share. Computing the Herfindahl index leads to a diversity index $2*0.45^2 + 0.1^2 = 0.415$. Now, suppose that an experienced resident enters the delivery room as a result of a request by one of the midwives. The resident is appreciated, has been around for a long time, and receives 80% of further intervention decision share. The midwives are left with 9% each, the nurse with 2%. This gives a diversity index of 0.657 as much decision power is now concentrated with the resident. It is closer to a monopoly. If the resident is a novice, however, who announces that he or she does not do cesarean sections and refuses to engage in other intervention decisions, then the resident's share could drop to say, 15%. This leaves the midwives with 40% each and the nurse 5%. Diversity increases as a result, with an index of 0.345.

But if the novice resident decides to call an experienced attending (who is well known in the hospital and well respected and therefore has a 90% share in intervention decisions), it could bring the index to 0.813, almost entirely monopolistic. Suppose that the resident is experienced, however, and that the attending and the rest of the team have just gone through resource management and nontechnical skills training with the aim to help downplay hierarchical boundaries. This would leave the attending with a 30% share in an intervention decision because he or she has learned how to take minority opinion seriously and empower nurses and midwives. The resident has 20%, the midwives 20% each, and the nurse 10%. H is now 0.22, the greatest diversity yet.

Complex systems can remain resilient if they retain diversity: The emergence of innovative strategies can be enhanced by ensuring diversity (Hollnagel et al., 2006). Diversity also begets diversity: With more inputs into problem assessment, more responses are generated, and new approaches can even grow as the combination of those inputs. As the example shows, however, diversity means not only adding team players but also making sure that there is a consideration of the weighting of their voices in any decisions that are made (or, in terms of complexity theory, different thresholds for when they speak up). In cases of drift into failure such the *Challenger* problem (recall from Chapter 5), unequal weighting of decision voices can be said to have played a role, with the company president of the contractor Morton Thiokol enjoying a near-monopoly on key decisions. Knowledgeable participants (e.g., the rocket scientists present in the telephone conference on the eve of the *Challenger* launch) have been depicted as having been held back by a very high threshold for speaking up (Feynman, 1988).

Systems that do not have enough diversity will be driven to pure exploitation of what they think they already know. Little else will be explored, and nothing new will be learned; existing knowledge will be used to drive through decisions. In business, this has been called a *takeover by dominant logic*, just as in politics it has been called groupthink. One of the positive-feedback loops that starts working with these phenomena is selection. People who adhere to the dominant logic, or who are really good at expressing the priorities and preferences of the team, will likely get heard.

This can legitimate senior practitioners who believe in the dominant logic, which offers even more incentives for subordinates to adhere to it as well.

The resilience of complex systems, however, derives not from compliance but from diversity: different practitioners who deploy differing and mutually sensitive repertoires for responding to evidence and to each other's constructions of, and concerns about, such evidence. Resilience is created when people keep a discussion about risk alive even when everything looks safe; teams and organizations have institutionalized the courage to show dissent independent of rank or status, to say "no" or "stop" even in the face of acute production pressures and when past successes are no longer seen as an automatic guarantee of future safety. Such creation of resilience applies across scales and levels—resilience can and must be created throughout hierarchies—not just at the sharp end.

There is, of course, a need for balance in the promotion of diversity. Too much diversity can mean that a team will keep on exploring new options and courses of action and never actually settle on one that exploits what it has already discovered and learned. In time-critical situations (such as in the delivery room), decisions may have to be taken without extensive exploration. Rather, the knowledge that is available there and then must be exploited even though better alternatives may lie just around the corner.

BRIEFINGS AND CHECKLISTS

Many safety-critical worlds do not attempt an operation without first briefing it thoroughly and then debriefing it when they are done. An operational briefing typically involves the following (Practicing Perfection Institute 2007):

- Reviewing the critical steps of the activity. This may include any steps that have immediate consequences or impact or those that if done incorrectly, could result in significant negative consequences.
- Anticipating error-likely situations. What are the potential error traps, and are there any conditions that could make falling into them more likely?
- Foreseeing potential consequences. Review what has gone wrong in previous operations of the same kind. What is the worst thing that could happen, and what is the most likely bad thing that could happen?
- Assessing the resilience in the team. Discuss the individual roles of the people who make up the operational team. Identify and if necessary, rehearse critical communications or callouts that nobody can afford to have misunderstood.

With some modification, these steps can also be used for debriefing after the operation is finished. As one form of preoperational briefing, many hospitals have instituted a so-called time-out, particularly before surgery. A main aim of such a time-out is to accentuate collective team responsibility for determining the correct site and side of surgery, but it does more than that. The time-out involves a final check of patient details, including surgical procedure and site, immediately prior to surgical preparation of the operative site. Ideally, a time-out should also confirm the roles of and expectations of each of the team members for that specific procedure and again reiterate the importance of speaking up when things are in motion.

The results of time-outs have been encouraging. In one hospital, a total of 10,330 procedures were performed during the 6 months after a time-out procedure was instituted. For these procedures, 9,098 (87.2%) completed time-out forms were returned and analyzed. There were no wrong-side or wrong-site surgeries performed during this 6-month period. However, there were three near-miss events that were captured by the time-out procedure. Analysis of the time-out forms also revealed numerous consent issues, incorrect documentation, and systems errors that could potentially have led to serious errors in clinical management. Although there were 109 objections (1.2%) to the time-out procedure during this initial period, the time-out gained acceptance among both medical and nursing staff as a valuable check prior to surgery (Hooper, Darley, Patton, Perry, & Skelton, 2006).

Time-outs are ideally part of a more complete protocol, like that developed by The Joint Commission, formerly the Joint Commission on Accreditation of Healthcare Organizations (JCAHO) in the United States, to prevent wrong-site and wrong-patient surgery. The protocol acknowledges that there are no effective single solutions to such complex problems. Robust fixes depend on multiple overlapping countermeasures and layers of protection (Wachter, 2008). They begin with preoperative verification of patient, procedure, and documentation; continue with marking the operative site in a way that will remain visible even after the patient has been prepped and draped; and culminate in the time-out that serves in part as a review of these previous processes just before surgery.

The need for standardization in these processes can be obvious. Although orthopedic surgeons had been marking surgical sites in the years before the protocol, a lack of standardization in such markings actually introduced additional risk. Some surgeons placed an X on the site or side to be operated on, while others marked sites or sides with an X that meant "do not cut here" (Wachter, 2008). Also, without a standardized approach to a time-out, people might not know what they should do or say during the time-out and will remain silent about concerns that (they think) they should resolve themselves. This is when a checklist can come in handy.

WHAT DOES A CHECKLIST DO?

A checklist functions not only as a basic memory guide but also as a generator and coordinator of operational work. Almost all checklists involve the following strategy, which can be done by one or multiple persons (Degani & Wiener, 1990):

- Reading or hearing the checklist item
- Accomplishing the item—either by verification of the correct setting or by execution of the checklist item
- Responding to the outcome of the action performed

A checklist can be seen, from a human factors perspective, as an additional interface between humans and the task or between humans and a machine. It helps control the method and sequence of work. It helps humans configure things technically and helps them mutually adjust and coordinate their own work and interplay. For this to work, the checklist needs to be well grounded in operational reality and not ask people to look at or do things when there is not time to do so. It also needs to abide by certain design guidelines itself.

A checklist ideally follows the flow of system settings or task steps in a logical manner. It intends to achieve the following objectives (Degani & Wiener, 1990):

- Aid the practitioner in recalling the process of configuring a device or remembering task steps without overlooking any of them
- Provide a standard foundation for verifying device configuration that will counteract any erosion in human psychological and physical functioning
- Provide convenient sequences for motor movements and eye fixations along the controls and displays of a device
- Provide a sequential set of steps to meet internal and external operational requirements that integrates other professions or team members
- Allow mutual supervision (or cross checking) among team members
- Enhance a team concept and involvement of all practitioners in configuring tasks and devices by keeping all team members in the loop
- Specify the duties of each crew member to facilitate optimum team coordination as well as logical distribution of workload
- Enhance coordination and smooth performance during stressful or high-workload conditions
- Reduce variability among practitioners so that they become more equivalent actors
- Serve as a quality control tool by management and regulators

Use of checklists basically comes in two varieties. The first is indeed to use the list as a "check" list. In this case, the checklist is used to go through a number of items *after* they have already been accomplished. Practitioners use their memory to set up a device or prep a patient and only then reach for the checklist and run through it to verify that critical items have not been forgotten and have been correctly accomplished. The second is to use the checklist as a "do" list. The list is then employed as a prompt for each of the actions to be taken in sequence. They are read off the list and then done—one by one. One problem with using a checklist as a do list is that it requires a practitioner to physically hold a list (although this may also be done by another practitioner) while manipulating other things in the work setting.

Responding to or verifying the checklist items can also be done in either of two ways, whether the list is used as a check or a do list. The first is known as *challenge-response*. This typically involves more than one person. One person reads the checklist items, and others either do or verify the items and report their status verbally (which is the response to the challenge). The second is known, among other labels, as *call-verify*. In this case, applying the list can involve only one person, who reads the item him- or herself and then does or checks it him- or herself. This can be done out loud or silently. Some of the redundancy of having a checklist called out in public is lost with this last method.

Which items should be on the checklist, and which can be left off? This is a vital question that, like any form of standardization, can intrude in beliefs about medical competence and practitioner identity (see Chapter 1). The application of the presurgical checklist (Haynes et al., 2009) in one hospital, for example, led to calls by senior physicians for the removal of the items "Confirm all team members have introduced themselves by name and role" and "Surgeon, anesthesia, and nurse verbally confirm patient, site, and procedure." They were not doing those items in practice anyway, and the reasoning went, if they had to confirm who they themselves were and who their patient was, then they had no business being in the operating theater in the first place. This is consistent with (particularly) surgical resistance against the introduction of the Safer Surgery checklist of the World Health Organization, as captured by Gawande (2010):

> We doctors remain a long way from actually embracing the idea. The checklist has arrived in our operating rooms mostly from the outside in and from the top down. It has come from finger-wagging health officials, who are regarded by surgeons as more or less the enemy, or from jug-eared hospital safety officers, who are about as beloved as the playground safety patrol. Sometimes it is the chief of surgery who brings it in, which means we complain under our breath rather than raise a holy tirade. But it is regarded as an irritation, an interference on our terrain. This is my patient. This is my operating room. And the way I carry out an operation is my business and my responsibility. So who do these people think they are, telling me what to do? (pp. 159–160)

The use of checklists is always about a balance between adapting and regimenting, between individual initiative and standardized sequencing. Experience and expertise are equally or even more important resources for action and may be distributed across multiple people in a team. Checklists are not meant to supplant human thinking but to support it. They also support interaction, coordination, and role articulation. They can actually help bring out human expertise and make sure that experiences from multiple team members are heard and brought to bear on the problem to be solved.

Checklists also are not ossified or static documents. In aviation, checklists are living documents that carry clearly posted issue numbers and publication dates. As operational experiences accumulate or change over time, checklists are amended accordingly and always in tight collaboration among equipment manufacturer, operator, regulator, and the practitioners who actually use them.

Checklists do not take the initiative to act or check out of human hands—no checklist can dictate its own application. The matching of context and checklist is still a deeply human endeavor that takes skill and situational judgment. So, checklists should not be seen as eroding medical heroism or practitioner identity. They can be seen as giving new, additional forms of expression to it.

How Should a Checklist Look?

In their 1983 study of human reliability, Swain and Guttman analyzed human error probabilities for various tasks in nuclear power plant operations. They recognized that a checklist without a check-off provision is more susceptible to errors of omission than a procedure with check-off provisions. The former could produce a per-time error rate of 1 in 100 against a much lower rate of only 3 in 1,000 if some kind of check-off provision were available (Degani, Heymann, & Shafto, 1999; Degani & Wiener, 1990).

Some checklists have this provision (e.g., check boxes or other hardware solutions that can slide down the list as items are accomplished), but many do not. As a result, practitioners employ different techniques to remember where they were on the list as they direct their attention to the setting in which the work is to be done or verified. For example, they may move a thumb along the items, but that only works if there is enough spacing between the items and if the lists are not distributed along multiple side-by-side columns. Some items also take a long time to complete, which forces the user to hold the list with the thumb in place for longer than is practical. If this is the case, perhaps the use of multiple checklists could be a solution (with a break where the longer item fits).

Then, there is the question of how to call things and how to call their setting or verification. Just asking practitioners to look at a items and to say "set" or "check" is not much of a verification. Compound checks should be avoided (as in the following: "IV lines—check") because one or more things may be missed even if some of the things in the compound check are verified. Specificity of what the item refers to is important, even if it means that things might take slightly longer to verify. Also, responses that simply say "set" or "check" are not useful. Proficiency and years of experience on type generates expectancies about how particular switches will be set and an automatic execution of actions related to the verification of their position, to the point that prerational, top-down influences will have an impact on what is seen, overriding bottom-up visual processing. The so-called mental model of the task and the environment in which it is to be executed can, after countless repetitions, become the stand-in for the actual execution and environment. Particularly, the case of an improper switch setting being verified as if it were proper is not at all uncommon (Degani and Wiener 1990, pp. 38–39):

> Many of the pilots interviewed by the authors stated that at one time or another they had seen a checklist item in the improper status, yet they perceived it as being in the correct status and replied accordingly. For example, the flap handle is at the zero degree slot (physical stimulus), but the pilot perceives its location on the 5 degree position, and calls "flaps—5," because he expects it to be there. This incorrect reply is based on numerous similar checks in which the flap handle was always in the proper setting during this stage in the checklist. ... Many times checklist procedures become an automatic routine ("sing-song" as some called it). The pilot would "run" the checklist, but the reply would be done from memory, and not based on the actual state of the item. The authors believe this is controlled by the output of the brain's pattern analyzing mechanism, and that the check procedure is done without conscious perception.

Conducting a checklist without conscious perception is as good as not doing a checklist at all. It is important that checklists be written and designed in ways that keep people looking at specific items in search for specific things to read out. Without that, attention can easily be diverted to other things, and people will begin responding in rote fashion just because they have done so before, not because they are seeing something that requires a response.

In an incident with parallels to hypoxic events in anesthesia, a Boeing 737 kept climbing on autopilot in 2005 while the cabin pressurization system was not working. This was not the first time that this happened in that particular aircraft, but in this case, both pilots became incapacitated and died. The aircraft crashed hours later when it had run out of fuel, killing everyone on board.

The switch that controls the pressurization system was likely set in manual because of maintenance work that had been carried out on the aircraft during the night before the accident flight. Although there is a checklist that covers the pressurization mode selector switch, it did not alert the pilots to a manual setting of the pressurization mode selector. The design of the checklist in fact violated various already established human factors principles, particularly those that warn against the use of compounds and underspecifications (Degani & Wiener, 1990):

Many checklists examined by the authors employ the ambiguous responses "set, check, completed," etc. to indicate that an item is accomplished. Instead, we believe that whenever possible, the response should always portray the actual status or the value of the item (switches, levers, lights, fuel quantities, etc.). (p. 41)

The preflight checklist told pilots to "verify Pack(s), Bleeds ON, SET." This was not only a compound check (combining the air conditioning system and the pressurization system), but also underspecified. Only the third confirmation (SET) referred to the pressurization panel but failed to address specifically any setting or switch. One of the things that needs to be done on the pressurization panel is selecting the (expected) cruise altitude and landing field elevation, which may typically be the main item that crews check when called to verify via their checklist that the air conditioning and pressurization panels are "SET." Having done this many times, such a check of the cruise altitude and landing elevation, in the absence of any clearer guidance or pointer from the checklist, may well come to stand for the complete check in response to the "SET" call on the checklist.

After takeoff, there was another, compound and underspecified check of the air conditioning and pressurization systems, identical to the preflight one. Item 1 of 4 on the after takeoff checklist asked pilots "AIR COND & PRESS ... SET." Repeating the same compound and underspecified question on a checklist that was not capable of catching a manual setting on the ground also did not help in the air. Also, in the phase directly after takeoff, there is a great possibility of the checklist being interrupted by, and having to be dovetailed with, other concurrent tasks, such as engine power settings, heading changes, radio frequency switches, routing updates as a result of new Air Traffic Control (ATC) clearances received after takeoff and so forth.

Checklists can be vulnerable to interruptions if the setting does not briefly insulate team members from what goes on around them (Loukopoulos, Dismukes, & Barshi, 2009). Other parties wanting things from the team, or team members joining in while a checklist is being read, can all be hugely disruptive. This can lead to things having to be redone or rechecked or to the need for reordering items to accommodate the requirements of the other practitioner or party. Such things, however, are always disruptive, and it might be safer to have a list than not having one—at least there is one unmistakable memory in the world of what needs to be done: the checklist.

KEY POINTS

- There are a number of human factors interventions that can be seen as practical tools for creating safety in healthcare—as they are in other industries. These include safety reporting and organizational learning, adverse event investigations, resource management training, and checklists.
- Effective reporting systems are nonpunitive, protected, and voluntary. Gathering reports is only the first step. Organizational learning is about the astuteness of the organization and the honesty and curiosity of its members in uncovering and analyzing problems. It can be seen as a collection of processes with which organizations improve their ability to accomplish their objectives by analyzing their past efforts.
- Getting people to report is about building trust: getting organizational members to realize that no adverse consequences come from them telling their stories. Keeping up the reporting rate is about participation and empowerment: involving organizational members in meaningful change.
- The ideal hospital safety department is keenly connected to the daily, messy details of healthcare practice, is independent from political or economic pressures inside and outside the organization, has immediate access to relevant managerial decision-making levels, and is constructively involved in organizational actions and change efforts that involve safety.
- An adverse event can in principle be told from an infinite number of perspectives, but the so-called first and second stories represent the bookends. In the first story, human error is the cause of the adverse event. Progress in safety can be made by protecting systems from unreliable people through selection, procedures, automation, training, and discipline. The second story sees human error as a symptom of trouble deeper inside the system, systematically connected to features of people's tools, tasks, and operating environment.
- The realization that most adverse events are triggered not by a lack of technical prowess or knowledge but by the breakdown of coordination between people (and people and technology) has been a watershed in safety thinking and has led to the development of checklists and resource management/ human factors training.

REFERENCES

American Medical Association. (1998). *A tale of two stories: Contrasting views of patient safety.* Chicago: Author.

Angell, I. O., & Straub, B. (1999). Rain-dancing with pseudo-science. *Cognition, Technology and Work, 1,* 179–196.

Argyris, C., & Schön, D. (1978). *Organizational learning: A theory of action perspective.* Reading, MA: Addison-Wesley.

Barach, P., & Small, S. D. (2000). Reporting and preventing medical mishaps: Lessons from non-medical near miss reporting systems. *British Medical Journal, 320*(7237), 759–763.

Billings, C. E. (1996). Situation awareness measurement and analysis: A commentary. In D. J. Garland & M. R. Endsley (Eds.), *Experimental analysis and measurement of situation awareness* (pp. 1–5). Daytona Beach, FL: Embry-Riddle Aeronautical University Press.

Billings, C. E. (1997). *Aviation automation: The search for a human-centered approach.* Mahwah, NJ: Erlbaum.

Columbia Accident Investigation Board. (2003). *Report Volume 1, August 2003.* Washington, DC: Author.

Cook, R. I., & Nemeth, C. P. (2010). "Those found responsible have been sacked": Some observations on the usefulness of error. *Cognition, Technology and Work, 12,* 87–93.

Cook, R. I., Nemeth, C., & Dekker, S. W. A. (2008). What went wrong at the Beatson Oncology Centre? In E. Hollnagel, C. P. Nemeth, & S. W. A. Dekker (Eds.), *Resilience engineering perspectives: Remaining sensitive to the possibility of failure.* Aldershot, UK: Ashgate.

Degani, A., Heymann, M., & Shafto, M. (1999). *Formal aspects of procedures: The problem of sequential correctness.* Paper presented at the 43rd annual meeting of the Human Factors and Ergonomics Society, September 1999 Houston, TX.

Degani, A., & Wiener, E. L. (1990). *Human factors of flight-deck checklists: The normal checklist.* (Contract Report NCC2-377) Moffett Field, CA: NASA Ames Research Center.

De Keyser, V., & Woods, D. D. (1990). Fixation errors: Failures to revise situation assessment in dynamic and risky systems. In A. G. Colombo & A. Saiz de Bustamante (Eds.), *System reliability assessment* (pp. 231–251). Dordrecht, the Netherlands: Kluwer Academic.

Dekker, S. W. A. (2002). Reconstructing the human contribution to accidents: The new view of human error and performance. *Journal of Safety Research, 33*(3), 371–385.

Dekker, S. W. A. (2007). Discontinuity and disaster: Gaps and the negotiation of culpability in medication delivery. *Journal of Law, Medicine and Ethics, 35*(3), 463–470.

Dekker, S. W. A., & Hugh, T. B. (2010). Balancing "no blame" with accountability in patient safety. *New England Journal of Medicine, 362*(3), 275.

Dekker, S. W. A., & Laursen, T. (2007). From punitive action to confidential reporting: A longitudinal study of organizational learning. *Patient Safety and Quality Healthcare, 5,* 50–56.

Feynman, R. P. (1988). *"What do you care what other people think?": Further adventures of a curious character.* New York: Norton.

Fischer, U., Orasanu, J. M., & Montvalo, M. (1993). *Efficient decision strategies on the flight deck.* Paper presented at the Seventh International Symposium on Aviation Psychology, April 1993 Columbus, OH.

Fischhoff, B. (1975). Hindsight ≠ foresight: The effect of outcome knowledge on judgment under uncertainty. *Journal of Experimental Psychology: Human Perception and Performance, 1*(3), 288–299.

Gawande, A. (2010). *The checklist manifesto: How to get things right.* New York: Metropolitan Books.

Gergen, K. J. (1999). *An invitation to social construction.* Thousand Oaks, CA: Sage.

Haynes, A. B., Weiser, T. G., Berry, W. R., Lipsitz, S. R., Breizat, A. H., Dellinger, E. P., et al. (2009). A surgical safety checklist to reduce morbidity and mortality in a global population. *New England Journal of Medicine, 360*(5), 491–499.

Hollnagel, E. (2009). *The ETTO principle: Efficiency-thoroughness trade-off. Why things that go right sometimes go wrong*. Aldershot, UK: Ashgate.

Hollnagel, E., Woods, D. D., & Leveson, N. G. (2006). *Resilience engineering: Concepts and precepts*. Aldershot, UK: Ashgate.

Hooper, G., Darley, D., Patton, D., Perry, A., & Skelton, R. (2006). Time-out, avoiding wrong site surgery: An audit of 6 months experience. *Journal of Bone and Joint Surgery, 88*(Suppl. 2), 311.

International Civil Aviation Organization. (1998). *Human factors training manual* (No. 9683-AN/950). Montreal: Author.

Janis, I. L. (1982). *Groupthink* (2nd ed.). Chicago: Houghton Mifflin.

Jensen, C. (1996). *No downlink: A dramatic narrative about the* Challenger *accident and our time*. New York: Farrar, Straus, Giroux.

Kohn, L. T., Corrigan, J., & Donaldson, M. S. (2000). *To err is human: Building a safer health system*. Washington, DC: National Academy Press.

Leape, L. L. (1994). Error in medicine. *Journal of the American Medical Association, 272*(23), 1851–1857.

Loukopoulos, L. D., Dismukes, K., & Barshi, I. (2009). *The multitasking myth: Handling complexity in real-world operations*. Farnham, UK: Ashgate.

Mahler, J. G. (2009). *Organizational learning at NASA: The* Challenger *and* Columbia *accidents*. Washington, DC: Georgetown University Press.

Nevile, M. (2002). Coordinating talk and non-talk activity in the airline cockpit. *Australian Review of Applied Linguistics, 25*(11), 131–146.

Nevile, M. (2004). *Beyond the black box: Talk-in-interaction in the airline cockpit*. Aldershot, UK: Ashgate.

Nevile, M., & Walker, M. B. (2005). *A context for error: Using conversation analysis to represent and analyse recorded voice data*. Canberra: Australian Transport Safety Bureau.

Orasanu, J. M. (1990). *Shared mental models and crew decision making* (No. CSL Report 46). Princeton, NJ: Cognitive Science Laboratory, Princeton University.

Orasanu, J. M., & Martin, L. (1998). Errors in aviation decision making: A factor in accidents and incidents. In *Human Error, Safety and Systems Development Workshop (HESSD) 1998*. Retrieved February 2008, from http://www.dcs.gla.ac.uk/~johnson/papers/seattle_hessd/judithlynnep

Page, S. E. (2007). Making the difference: Applying a logic of diversity. *Academy of Management Perspectives, 11*, 6–20.

Practicing Perfection Institute. (2007). *The pre-job brief*. Swanzey, NH: Author.

Predmore, S. C. (1991). *Microcoding of communications in accident investigation: Crew coordination in United 811 and United 232*. Paper presented at the Sixth International Symposium of Aviation Psychology, Columbus, OH.

Pronovost, P. J., & Vohr, E. (2010). *Safe patients, smart hospitals*. New York: Hudson Street Press.

Reason, J. T. (1990). *Human error*. New York: Cambridge University Press.

Reason, J. T. (1997). *Managing the risks of organizational accidents*. Aldershot, UK: Ashgate.

Rochlin, G. I. (1999). Safe operation as a social construct. *Ergonomics, 42*(11), 1549–1560.

Rochlin, G. I., LaPorte, T. R., & Roberts, K. H. (1987). The self-designing high reliability organization: Aircraft carrier flight operations at sea. *Naval War College Review*, 76–90.

Rogers, W. P., et al. (1986). *Report of the Presidential Commission on the Space Shuttle* Challenger *Accident*. Washington, DC.

Sagan, S. D. (1993). *The limits of safety: Organizations, accidents, and nuclear weapons*. Princeton, NJ: Princeton University Press.

Salas, E., Wilson, K. A., & Burke, C. S. (2006). Does crew resource management training work? An update, an extension, and some critical needs. *Human Factors, 48*(2),
392–413.

Schein, E. (1992). *Organizational culture and leadership* (2nd ed.). San Francisco: Jossey-Bass.

Starbuck, W. H., & Farjoun, M. (2005). *Organization at the limit: Lessons from the* Columbia
disaster. Malden, MA: Blackwell.

Svenmarck, P., & Dekker, S. W. A. (2003). Decision support in fighter aircraft: From expert
systems to cognitive modelling. *Behaviour and Information Technology, 22*(3), 175–185.

Vaughan, D. (1996). *The* Challenger *launch decision: Risky technology, culture, and deviance
at NASA.* Chicago: University of Chicago Press.

Vaughan, D. (1999). The dark side of organizations: Mistake, misconduct, and disaster. *Annual
Review of Sociology, 25,* 271–305.

Wachter, R. M. (2008). *Understanding patient safety.* New York: McGraw-Hill.

Weick, K. E. (1988). Enacted sense-making in crisis situations. *Journal of Management
Studies, 25,* 305–317.

Weick, K. E. (1990). The vulnerable system: An analysis of the Tenerife air disaster. *Journal
of Management, 16*(3), 571–594.

Wildavsky, A. B. (1988). *Searching for safety.* New Brunswick, NJ: Transaction Books.

Woods, D. D. (1993). Process-tracing methods for the study of cognition outside of the
experimental laboratory. In G. A. Klein, J. M. Orasanu, R. Calderwood, & C. E.
Zsambok (Eds.), *Decision making in action: Models and methods* (pp. 228–251).
Norwood, NJ: Ablex.

Woods, D. D. (2006). How to design a safety organization: Test case for resilience engineering. In E. Hollnagel, D. D. Woods, & N. G. Leveson (Eds.), *Resilience engineering:
Concepts and precepts* (pp. 296–306). Aldershot, UK: Ashgate.

Woods, D. D., Dekker, S. W. A., Cook, R. I., Johannesen, L. J., & Sarter, N. B. (2010). *Behind
human error.* Aldershot, UK: Ashgate.

Xiao, Y., & Vicente, K. J. (2000). A framework for epistemological analysis in empirical (laboratory and field) studies. *Human Factors, 42*(1), 87–102.

Zaccaria, A. (2002, November 7). Malpractice insurance crisis in New Jersey. *Atlantic Highlands
Herald.* http://www.ahherald.com/physicians_forum/2002/pf021107_malpractice.htm

7 Accountability and Learning from Failure

The single greatest impediment to error prevention is that we punish people for making mistakes.

**Lucian Leape, in testimony on Veterans Affairs
before the U.S. Congress, 1999**

LEARNING AND ACCOUNTABILITY—JUST CULTURE

For most people, it is not hard to agree on the basics. Attributing adverse events to human error does little to improve the healthcare system. If we remove the error, or the individual who made the error, all we do is create an illusion of progress on safety. It denies the existence of system issues that gave rise to the error or its potential. It simply identifies a scapegoat, and it truncates learning anything meaningful from the event. Blame inhibits learning. And simple attributions of an adverse event to "human error" stop deeper investigation and hamper understanding (Dekker, 2006; Morath & Turnbull, 2005; Woods, Dekker, Cook, Johannesen, & Sarter, 2010). It is not difficult to understand that a system of adverse event reporting and investigation needs to take a blameless approach if it wants to have the organization learn anything of value.

A learning culture is a culture that allows the boss to hear bad news (Dekker, 2002). Effective risk management and organizational improvement depend on open communication about safety incidents (Institute of Medicine, 2003) and on accepting that errors are the inevitable by-product of pursuing success under the ordinary organizational conditions of resource scarcity and goal conflicts (Woods et al., 2010). Errors and adverse events come from somewhere—there are conditions that make their recurrence likely. Learning about those conditions and doing something about them is a good investment in safety. As Mahler (2009) put it:

> In a learning organization, problems are openly acknowledged, causes are intensively investigated, and procedures are corrected. New techniques are searched out, all in order to bring results more closely in line with expectations, external mandates and professional or personal values. (p. 18)

Again, those are the basics. Create a culture that allows bad news to get to the boss or a culture in which the boss wants bad news, demands it, expects it. But not all bad news is equal, is it? Are some errors worse or more blameworthy than others?

Are certain actions more at risk or reckless than other behavior or others' behavior? Is there not simply some behavior that actually is culpable, if anything to make an example out of it so that other people might think twice before doing a similar thing? These are questions that managers and colleagues in healthcare frequently raise.

The immediate dilemma for a boss is that he or she wants to hear all the bad news but cannot accept all the bad news. Yet if the boss is clear in letting an employee know that the bad news is unacceptably bad (e.g., through a punitive response), then chances are that bad news will no longer travel to the boss.

From the position of a manager or administrator, the best way to manage this balance is to involve the practitioners who would potentially report (not necessarily the one who *did* report because if it is a good reporting system, that might not be known to the manager). What is their assessment of the "error" that was reported? How would they want to deal with a colleague who did such a thing? In the end (and as discussed in more detail in this chapter), whether an error is culpable is not about the crossing of some clear line that was there before the error. Perceived justness of a response to a reported problem sometimes lies less in the response than in who is involved in concocting that response.

For a manager, keeping the dialogue open with his or her practitioner constituency must be the most important aim. If dialogue is killed by rapid punitive action, then a version of the dialogue will surely continue elsewhere (behind the back of the manager). That leaves the system none the wiser about what goes on and what should be learned.

Responses to bad news matter. As Weiner and colleagues pointed out (Weiner, Hobgood, & Lewis, 2008), the term *just culture* is often used narrowly to refer to the beliefs, assumptions, and expectations that govern disciplinary responses to unsafe acts inside an organization. A just culture is meant to offer a nonpunitive environment in which individuals can report errors or close calls without fear of reprimand, rebuke, or reprisal, yet it does not offer an environment in which no accountability exists.

Accountability is seen by many here as just that: the response to any controversial or potentially blameworthy act. Accountability is, however, a much more complex social construct, fundamental to human relationships (Dekker, 2007d; Lerner & Tetlock, 1999). The problem of making the culture just goes both ways. If an honest mistake is punished, people will see this as an unjust response. If reckless behavior is left unpunished, people also will see this as unjust. What matters is that people in healthcare feel assured that they will receive fair treatment when they are involved in adverse events or report them. But how can we get this right?

JUST ASSIGN BEHAVIOR TO THE RIGHT CATEGORY (RIGHT?)

The primitive belief in healthcare is that all we need to do is distinguish among human error, at-risk behavior, and reckless behavior (Marx, 2001). Once we know in which of these categories an action falls, we have a good idea of how to respond to it. The distinctions are assumed to matter because they imply a steadily more aggravated role played by people's bad intentions. Human error is seen as the mildest form. It is an inadvertent action over which people have no control—neither over the error nor over its outcome. At-risk behavior has a larger intentional component.

It consists of knowingly being noncompliant. Different people can have different ideas about the behavior and its trade-off between risk and reward, but in the end, it is likely to invite some type of sanction or other consequence. Reckless behavior, however, ranks as the most conscious and typically deserves a sterner response. It is described as a choice to consciously disregard a substantial and unjustifiable risk.

The model is of course familiar to anybody in healthcare and as such already vaguely reassuring. It is like matching symptoms to a disease. If the symptoms of the behavior are recurring (i.e., people do this bad thing all the time) or unique to this person (as opposed to the group of practitioners as a whole), then it would seem that this individual is engaging in reckless (as opposed to at-risk or erroneous) behavior.

But categorizing other people's behavior, particularly on the basis of assumptions about their intentions, consciousness, and choices, is only apparently simple. In reality, it is deviously treacherous. Multiple factors conspire against our ability to fairly and meaningfully adjudicate what other people did or intended. The first factor is relatively simple. Recall from Chapter 6 (its discussion of categories in reporting systems) that as soon as we put down lines between different categories of behavior, we create a problem of boundaries. When does an error become at-risk behavior, or for that matter, reckless? Where exactly does an act cross that line? If the same act can fall into both categories at the same time, then having those categories is not useful. The whole point of categorization is that acts are uniquely one or the other because it is that assignment that lays out legitimate responses or countermeasures (Dekker, 2007c). If only it were that simple.

Attempts to map situated and often invisible aspects of human expertise such as decision making, insight, awareness, or consciousness onto discrete categories will always be a leap of faith—our own faith that we have it right. How do we know for sure whether something was deliberated or done consciously or intentionally? Even if we feel confident that we have some idea about that, these things are always a matter of degree (how much did the person in question really know or foresee?) and typically resist categorical answers.

The biggest problem with attempts at categorizing behavior lies with ourselves. It is our belief that human error, at-risk behavior, or recklessness are stable empirical categories of human performance (Cook & Nemeth, 2010). Also, the recklessness, risks, or erroneousness of those acts form an inherent property of other people's performance, and our own interpretations have nothing to do with us seeing that performance as such. But saying that others' behavior is erroneous, risky, or reckless is *our* judgment of what other people do, not a description of the essence of their behavior (Becker, 1963). In other words, if we categorize behavior, we do nothing more than categorize our own judgments. This will become even more obvious when we consider a definition of *negligence* in this chapter.

ASSIGNING BEHAVIOR TO CATEGORIES IS ABOUT POWER, PRODUCTION, AND PROTECTION

Just responses to adverse events are not a matter of matching the inherent properties of undesirable behavior with appropriate pigeonholing and a fitting punitive level. Many people in healthcare seem to think, however, that is exactly what just responses (and the creation of a just culture) are about (Marx, 2001). The reason that it does not work (or that such guidance circumvents the really hard problem in creating a just culture) is that those categories are nothing more than our own judgments, our own attributions. We still need to do the hard work of deciding whether to see something as reckless, as at risk, or as erroneous. And because this will forever be a judgment, it will forever be contestable by those who judge things differently.

Assigning acts to categories is a matter of power. Who has the power to say that behavior is one thing and not the other? And who has the power to decide on the response? As soon as power is involved, though, categorizations and responses may quickly be seen as unjust, as unfair. Suppose nurses take to scanning the bar code label that one of their colleagues pasted on the wall behind the patient because it actually reads well and is always easy to find as opposed to others. This may have become all-but-normal practice—everybody does it because everybody always has the next patient, and the next, and the medication bar code scanners are of such poor quality that they do not do a good job of reading anything except labels that are entirely flat and high contrast (which a label pasted on a wall is).

Managers may want to call such behavior at risk and mete out supposedly appropriate countermeasures. But nurses on the wards may no longer see their behavior as at risk, if they ever did. In fact, it may be a sure way to get a good scan and not doing that could create more risks. Of course, bar code scanning is not their main job—taking care of patients is. They have a job to do (in fact, they have many jobs to do), and this is a really good and reliable way of getting that job done. This means that nurses may see any punitive responses to scanning a label stuck on a wall as unjust. After all, such responses may show that the manager is not aware of the unrelenting pressures and ebbing and flowing demands of nursing work or of the shortcomings of putatively laborsaving and safety-enhancing technology. Justice is a matter of perception.

But managers are under different pressures. Managers appropriate the power to call something at risk not because of their privileged insight into the risks and realities of nursing work but because they *can* and because they *have to* relative to the pressures and expectations of their position in the hospital. Judgments about nurses' performance have to make local sense to the managers; otherwise, they would not make them. From a manager's point of view, operational behaviors that bypass instructions or protocol, for example, could end up eroding productivity and reputation and eventually impair the financial performance of their part of the organization. Or, for that matter, having to make structural or equipment changes (e.g., procuring new or better bar code scanners) involves sacrifices that are much larger than reminding people to be more careful and to follow the rules.

A patient being prepared for spinal surgery was placed on a modular table (Cook & Nemeth, 2010). One member of the surgical team noticed that there was a slight tilt to the table and began to correct the position of the table. The table swung loose, and the anesthetized patient fell from the table to the floor but sustained no injury.

The table is able to swivel around a centerline—something that is essential to position a patient for spinal surgery. It also presented a hazard because the ability of the table to rotate freely made it possible for an anesthetized patient to fall.

No member of the surgical team was ever assigned to be responsible for operating the table. It was a job that easily fell between the cracks of patient preparation and handling.

The table had to be operated through a rather complex and confusing set of displays, lights, and handles. But the surgical team really liked its positional versatility and insisted on using it, despite its risks. What could managers do? Ask people to try harder, of course, was an easy way forward, which is essentially what happened: Team members were cautioned about the risk ("A human error happened; do not let it happen to you"). The table or its controls were not modified. No warning sign was hung on the lever that made the table swivel; no regulatory limitations were introduced. Nothing was done.

The unit, as most units like it, was running at or near its saturation point, and there were simply no people, no time, and no resources to make changes that are more useful. The judgment that this was a human error simply produced too many organizational benefits.

Calling something erroneous, at risk, or reckless offers the idea of managerial control. It offers the illusion that such a behavioral label can eventually make the behavior stop. This is an illusion indeed. What keeps behaviors alive is not the label anybody gives to them, but the usefulness and rewards they offer to those who engage in them. And the usefulness and rewards, in turn, are bred by the circumstances in which, and the tools with which, such work is done. Slapping another label on it, thereby making it less legitimate, will not make it go away.

Research, however, has suggested that categorizations are not just about illusory control or the exercise of power. Psychological mechanisms of defense and distancing may operate as well, even, or especially, among colleagues. Colleagues might find that calling somebody else's behavior erroneous or at risk comes as a relief to them, as something that seems to make good sense. As Cook and Nemeth (2010) pointed out:

Pejorative qualities that are often attached to human error promote distancing, such as suggestions that error arises from sloth or moral failing. Others feel less at risk if error can be ascribed to a practitioner's deeply seated, but personal, flaws. If accidents arise from forces and circumstances in the environment, then the experience of my colleague has relevance for me and the event increases my sense of hazard and uncertainty. By attributing my colleague's accident to his inattention or stupidity, though, I make it possible to believe that the accident has no relevance for me. This is because I do not believe that I am either inattentive or stupid. (p. 91)

The egregiousness (or stupidity) of an act, then, is not something that is an essential property of the act. It is socially constructed by those who look at the act and who have to do something with it—respond to it, learn from it, protect themselves from it, deny that it may happen to them. That does not mean that there are no regularities in how people construct the culpability of acts, however. Assumptions of other people's volitional behavior and outcome control, as well as causal control, play a dominant role (Alicke, 2000). Such regularities have even been exploited in the construction of decision trees that can walk people through a set of questions to help them decide the blameworthiness of an act (Reason, 1997). Yet the questions offered confirm the negotiability of the line rather than resolving its location:

- Were the actions and consequences as intended? This invokes the judicial idea of a mens rea ("guilty mind") and seems a simple enough question. Few people in healthcare intend to inflict harm, although that does not prevent them from being prosecuted for their "errors" (under charges of manslaughter, for example, or general risk statutes that hail from road traffic laws on endangering other people). Also, what exactly is intent, and how do you prove it? And who gets to prove this, using what kind of expertise?
- Did the person knowingly violate safe operating procedures? People in operational worlds knowingly adapt written guidance and have to do so to bridge the gap between prescriptive routines and actual work in worlds of imperfect knowledge, time constraints, and infinite variation. Calling such adaptations "violations" already implies amoral judgment about who is wrong (the worker) and who is right (the rule). It is easy to show in hindsight which guidelines would have been applicable, available, workable, and correct for a particular diagnosis or clinical procedure (says who, though?), but such overestimations of the role of noncompliance in the wake of incidents conceals the real operational dilemmas faced by practitioners.
- Were there deficiencies in training or selection? The term *deficiencies* seems unproblematic, but what is a deficiency from one angle can be perfectly normal or even above industry standard from another. Indeed, the use of *deficiency* is not itself a judgment about the content, relevance, applicability, or any other feature of training or selection.

Thus, in trying to assign behavior to categories, questions such as these may form a good start. But they cannot arbitrate between culpable or blameless behavior. The line, after all, is not a location that can be written in a rule, procedure, or guidance document. It is a judgment that people will have to make again, every time.

ACHIEVING ORGANIZATIONAL JUSTICE

If we cannot simply say that some behavior *is* reckless, at risk, or erroneous, and if saying so says more about our power to render such a judgment than about the behavior in question, then where does that leave us? Most healthcare leaders actually have a keen interest in establishing a just culture in their organizations (Weiner et al., 2008) and may reach for anything that gives them the illusion of doing so, even if it leaves them holding the bag with all the hard interpretive work and the responsibility for making those sheer impossible judgments about other people's consciousness or intentions. The goal is to achieve organizational justice, but how can that be done meaningfully?

Justice is a perception. But whether something gets perceived as just depends on more than only the point of view of the perceiver. The outcome or response (and whether it gets construed as fair or not from that viewpoint) is of course important. The way of getting to the response also matters. This involves a number of things. What are the procedures used to determine the response? Who is involved in determining it? And is the person who receives the judgment involved and how? There are a few generic ways that will be seen as more just than others.

First, the persons involved in determining the response have to have some idea about the actual work they are discussing. They should have some knowledge of what it is to be there (to work as a nurse on a night shift with responsibility for 24 patients, for example). In other words, they should be part of, or at least be credibly representative of, the peer group.

Second, offering people inside an organization some kind of democratic process over who gets to produce these responses is a good way of having them seen as more just. Who has the power to assign behavior to these categories of erroneous, at risk, or reckless? Giving practitioners influence over those things will certainly go a long way toward building a just culture. The important internal discussion for an organization to have is who gets to draw the line between acceptable and unacceptable behavior. This means not only who gets to handle the immediate aftermath of an incident (the line organization supervisor/manager or a staff organization such as safety department) but also how to integrate practitioner or peer expertise in the decision on how to handle this aftermath.

A problem here is that the perceived seriousness of an event may change who gets to draw the line. In other words, any arrangements that are struck here cannot be kept entirely stable. The more serious an event is seen to be, the more likely it is that people higher up and even outside the organization will get involved and have their say about whether people's behavior transgressed some line of culpability. Yet being honest about this to practitioners is always an option—a good option at that, even if such honesty implies having to say that you do not really know who will eventually draw the line.

Third, if a response is produced without ever hearing the person to whom it is directed, that will likely be seen as unjust. Empowering and involving the practitioner him- or herself in the aftermath of an incident is the best way to maintain morale, maximize learning, and reinforce the basis for a just culture. Also, if the judgment or the response is balanced by other organizational measures (e.g., adequate peer support that helps people deal with them being the second victim of an adverse event; see the end of this chapter), then this may mitigate any negative reading of the response. People really care about how they are treated in the wake of an adverse event.

There are a number of additional measures that an organization or manager can take to foster a just culture (Dekker, 2009):

- An incident must not be seen as a failure or a crisis, neither by management nor by colleagues. An incident is a free lesson, a great opportunity to focus attention and to learn collectively.
- Abolish financial and professional penalties (e.g., suspension) in the wake of an occurrence. These measures render incidents as something shameful,

to be kept concealed, leading to the loss of much potential safety information and lack of trust.

- Monitor and try to prevent stigmatization of practitioners involved in an incident. They should not be seen as a failure or as a liability to work with by their colleagues.
- Implement, or review the effectiveness of, any debriefing programs or critical incident/stress management programs the organization may have in place to help practitioners after incidents. Such debriefings and support form a crucial ingredient in helping practitioners see that errors and incidents are normal (if undesirable) outcomes of work in healthcare, but that they can help the organization get better, and that they can happen to everybody.
- Build a staff safety department, not part of the line organization, that deals with incidents. The direct manager (supervisor) of the practitioner should not necessarily be the one who is the first to handle the practitioner in the wake of an incident. Aim to decouple an incident from what may look like a performance review or punitive retraining of the practitioner involved.
- Start with building a just culture at the beginning: during basic education and training of the profession. Make trainees aware of the importance of reporting incidents for a learning culture and have them see that incidents are not something individual or shameful but a good piece of systemic information for the entire organization. Convince new practitioners that the difference between a safe and an unsafe organization lies not in how many incidents it has but in how it deals with the incidents that it has its people report.
- Ensure that practitioners know their rights and duties in relation to incidents. Make as clear as you can what can (and typically does) happen in the wake of an incident (e.g., to whom practitioners were obliged to speak and to whom not). A reduction in such uncertainty can prevent practitioners from withholding valuable incident information because of misguided fears or anxieties.

REPORTING VERSUS DISCLOSURE

Finally, hospitals should think explicitly (if they do not do so already) about how to protect their data from undue outside probing (e.g., by a prosecutor). The consequences of this step must be considered carefully. One problem is that better protection for incident reporters can lock up information even for those who rightfully want access to it and who have no vindictive intentions (e.g., patients or their families). The protection of reporting can make disclosure to such parties more difficult.

The Hippocratic principles of beneficence and nonmaleficence ethically obligate healthcare workers to disclose their role in adverse events—owing to their unique, fiduciary relationship with patients. Disclosure is seen as a marker of professionalism. Contributing to organizational learning through event reporting creates an additional ethical commitment: in this case, not helping the patient but the organization (e.g., colleagues) understand what went wrong and how to prevent recurrence. Reporting also is increasingly seen as part of professionalism: drawing on the unique insights from sharp-end practice, from daily contact with safety-critical procedures and technologies, to help one's system become better and safer. But efforts to promote reporting may sometimes complicate the fulfillment of the obligation to disclose. They present a potential moral dilemma to consider when implementing event reporting and investigation programs.

Disclosure is the provision of information to patients and families. *Reporting* is the provision of information to supervisors, oversight bodies, or other agencies. Efforts to encourage reporting and disclosure could conflict. Reported events can lead to meaningful investigation and intervention, but they do not necessarily end up elucidating patients or families. In fact, protection of reporters (to encourage reporting) may require that the reported information somehow remains confidential, precluding disclosure (at least of that exact information). Conversely, disclosure may lead to legal or other adverse actions, which in turn can dampen people's willingness to either report or disclose. In the end, caregivers are accountable to both their organization (to help it learn and improve) and their patients (as a fundamental part of the fiduciary relationship). Such dual accountability perhaps means that different stakeholders need different kinds of stories: one that offers the organization insight into existing vulnerabilities and leverage for change, another that helps a caregiver discharge the professional responsibility to communicate what went wrong to those affected.

A Discretionary Space for Accountability

Moves to redirect the power to draw the line away from a manager or supervisor (or the judiciary for that matter; see the section on criminalization) can be met with suspicions that practitioners want to blame "the system" when things go wrong, and that they do not want to be held liable in the same way as other members of the organization or society. Yet perhaps the choice is not between blaming people or systems. Instead, we may reconsider the accountability relationships of people in systems.

All safety-critical work is ultimately channeled through relationships between human beings or through direct contact of people with the risky technology. At this sharp end, there is almost always a discretionary space into which no system improvement can completely reach (Sharpe, 2003). This is in part a space in the almost literal sense of "room for maneuvering" that operators enjoy while executing their work relatively unsupervised (in the examination room, the operating theater, cockpit). It is also a space in a metaphorical sense, of course, as its outlines are not stipulated by decree or regulation but drawn by actions of individual operators and the responses to them. It is, however, a final kind of space filled with ambiguity, uncertainty, and moral choices. And it is a space typically devoid of relevant or applicable guidance from the surrounding organization, leaving the difficult calls up to the individual practitioners. Systems cannot substitute the responsibility borne by individuals within that space. Individuals who work in healthcare would not even want that. The freedom (and concomitant responsibility) that is left for them is what makes them and their work human, meaningful, a source of pride.

But organizations can do a number of things. One is to be as clear as possible about where that discretionary space begins and ends. Not giving practitioners sufficient authority to decide on courses of action, but demanding that they be held accountable for the consequences anyway, creates impossible and unfair goal conflicts (for which managers may sometimes be held accountable, but they also could have been the recipients of similar goal conflicts). It effectively shrinks the discretionary space before action but opens it wide after any bad consequences of action become apparent.

Second, an organization must deliberate how it will motivate people to carry out their duties conscientiously inside the discretionary space. Is the source for that motivation fear or empowerment? There is evidence that empowering people to affect their work conditions, to involve them in the outlines and content of that discretionary space, most actively promotes their willingness to shoulder their responsibilities inside it.

> During surgery, an anesthetist reached into a drawer that contained two vials that were side by side, both with yellow labels and yellow caps. One, however, had a paralytic agent, the other a reversal agent for when paralysis was no longer needed. At the beginning of the procedure, the anesthetist administered the paralyzing agent. But toward the end, he grabbed the wrong vial, administering additional paralytic agent instead of its reversal agent. There was no bad outcome in this case. But when he discussed the event with his colleagues, he found that this had happened to them also, and that they were all aware of the potential risks. Yet none had spoken out about it, which could raise questions about the empowerment anesthetists may have felt to influence their work conditions, their discretionary space (Morreim, 2004).

ACCOUNTABILITY FREE IS NOT BLAME FREE

The conflation of accountability with responding to an adverse event creates a peculiar red herring. Equating blame-free systems with an absence of personal accountability, as some in healthcare do (Pellegrino, 2004), is misleading. The kind of accountability wrung out of practitioners in a trial or in front of a manager who holds all the cards is not likely to contribute to justice or to future safety in their field and in fact may hamper it.

We can create accountability not by blaming people, but by getting people actively involved in the creation of a better system to work in. Holding people accountable and blaming people are two quite different things. Blaming people may in fact make them less accountable: They will tell fewer accounts; they may feel less compelled to have their voice heard, to participate in improvement efforts. Blame-free or no-fault systems are not accountability-free systems. On the contrary, such systems want to open up the ability for people to hold their account so that everybody can respond and take responsibility for doing something about the problem.

This again has implications for what we mean by accountability. If we see an act as at-risk or reckless behavior (or as a crime), then accountability means blaming and punishing somebody for it. Accountability in that case is backward looking, retributive. If, instead, we see the act as an indication of an organizational, operational, technical, educational, or political issue, then accountability can become forward looking (Sharpe, 2004). The questions become this: What should we do about the problem, and who should bear responsibility for implementing those changes?

CRIMINALIZATION OF MEDICAL ERROR: A GROWING PROBLEM?

People in healthcare in various countries around the world are seeing an increase in the criminalization of medical error (Dekker, 2007a; Pandit, 2009). The laws under which criminal prosecution of professionals currently occurs are often derived by extending general hazard statutes, particularly from road traffic laws that criminalize the reckless endangerment of other people or property (Tingvall & Lie, 2010). In other cases, "long-shot" approaches may be taken such as the prosecution of a nurse's mistake on the basis of Medicare antifraud laws (Dekker, 2010). The move to criminalize human error (a label that is itself a psychological attribution; see Chapter 2) could parallel the evolution of, for example, law on hate crime, which went from a broad, ambiguous category to a focused, determinate legal construct. Indeed, Sweden has deliberated implementing the legal category "patient safety crime" (Jacobs & Henry, 1996; Phillips & Grattet, 2000).

Doubts have been raised about the fairness of criminalizing errors that are made in the course of executing normal professional duties with no criminal intent (Mee, 2007; Merry & Peck, 1995; Moran, 2008; Reissner, 2009) and the capriciousness of criminal prosecution (e.g., one nurse was criminally convicted for a medication administration error of a kind that was reported to the regulator by others more than 300 times that year alone; Ödegård, 2007). Doubts also exist about the ability of a judiciary to make sense of the messy details of practice in a safety-critical domain like healthcare (Anderson, 2005), let alone resist common biases of outcome knowledge and hindsight in adjudicating people's performance (see Chapter 2) (Hugh & Dekker, 2009).

Despite these concerns, there is no coherent program of research into the social causes of a trend toward criminalization in healthcare. Communities specializing in disciplines concerned with criminalization and victimization are segregated from those working on risk and safety. Interesting tensions and affinities across relevant work are hardly visible, and theoretical matters for debate have not been identified; a dialogue essential to intellectual development has not really started. This part of this chapter tries to fill part of the gap.

CRIMES AS INHERENTLY REAL OR AS CONSTRUCTED PHENOMENA

A broader theoretical issue is at stake here, as there is with the characterization of "human error" and at-risk and reckless behavior. In a field such as medicine, with its often positivist, technology-oriented, and androcentric biases, the nature of culpable acts is easily taken as essential and unproblematic (Bosk, 2003). Remember, practitioners in healthcare have (Leape, 1994) "come to view an error as a failure of character—you weren't careful enough, you didn't try hard enough. This kind of thinking lies behind a common reaction by physicians: 'How can there be an error without negligence?'" (p. 1851).

Such an epistemology is hostile to characterizations of error and negligence (or other kinds of crime) as relative, historically located, and observer-contingent constructions of perspective, background, and language. Indeed, some in healthcare reject this explicitly. "All negligent error," said Edmund Pellegrino (2004) about adverse medical events, "is morally blameworthy" (p. 86). Negligent error, in such a reading, is taken to exist out there, independently of the observer, unproblematically, ready formed. If healthcare professionals make errors that are negligent, they should be considered morally blameworthy.

This feature of healthcare is consistent with how criminology has long adhered to a fairly narrow scientific essentialism that sees social facts as inert and stable across observers and observations (Bjarup, 2005; Rafter, 1990). "Criminal" aspects of mistakes are seen as nonarbitrary empirical facts that are dealt with by the legitimated authorities (North, 2000). The idea is that if there is evidence of negligence or recklessness, then that behavior needs to be dealt with as such, but by those people who have the authority to do so. This leaves little room for critical reflection on who constructed the alleged act as reckless or negligent and from which political, managerial, organizational, or other social force field this construction emerged (Merton, 1938; Summerton & Berner, 2003).

The resulting theoretical position may have sacrificed engagement with the criminalization of mistake as a social-scientific issue in healthcare. But that which is supposedly "criminal" is not clear or essentially obvious, in spite of all theoretical commitments and professional inculcation to the contrary. Take this definition of negligence, for example:

> Negligence is conduct that falls below the standard required as normal in the community. It applies to a person who fails to use the reasonable level of skill expected of a person engaged in that particular activity, whether by omitting to do something that a prudent and reasonable person would do in the circumstances or by doing something that no prudent or reasonable person would have done in the circumstances. To raise a question of negligence, there needs to be a duty of care on the person, and harm must be caused by the negligent action. In other words, where there is a duty to exercise care, reasonable care must be taken to avoid acts or omissions which can reasonably be foreseen to be likely to cause harm to persons or property. If, as a result of a failure to act in this reasonably skillful way, harm/injury/damage is caused to a person or property, the person whose action caused the harm is negligent. (GAIN 2004, p. 3)

This definition does not provide any answers regarding what negligence is. Rather, it lays out an array of new questions and judgments that someone must make. What is normal standard? How far is below? What is reasonably skilful? What is reasonable care? What is prudent? Was harm indeed caused by the negligent action? It is not that we cannot, in principle, find answers to these questions, but no definition of negligence captures the essential properties of negligence, so that we could grab negligent behavior and put it on the unacceptable side of some line.

It is seductive, of course, to think that once we weed through the questions surrounding an unwanted act and its negative consequences, we can "really" discover whether there is negligence behind it, as if we were following just another clinical guideline (indeed, much existing guidance on just culture sells precisely this false idea; Marx, 2001). But the belief in a good method or procedure that will lead us to the essence of an act is sustained in court as well. We believe that courts can tease out that reality, that truth. But the position taken by most of those in social science over the last 30 or 40 years is that there is nothing inherently "true" about the error or its negligent nature. Its meaning is produced, enforced, and handed down through social and professional systems of language and institutions (Becker, 1963, p. 9):

> Deviance is created by society. Social groups create deviance by making the rules whose infraction constitutes deviance and by applying those rules to particular persons and labeling them as outsiders. From this point of view, deviance is not a quality of the act the person commits, but rather a consequence of the application by others of rules and sanctions to an "offender." The deviant is the one to whom the label has success-fully been applied; deviant behaviour is behaviour that people so label.

What counts as negligent is the outcome of processes of societal negotiation, of social construction, by which an act is turned into an error and the error into negligence. Consider what this means for putting somebody on trial for medical error. The question asked frames the search for and interpretation of findings: Did this error amount to a crime? Remember that the notion of error is already deeply troublesome—a negotiated construction or psychological attribution rather than a simple, observable fact in reality. In judging whether a medical error is a crime, then, we try to see whether one social construct can be construed as another. Just as the properties of an error are not objective and independently existing (see Chapter 2), so a crime arises from our ways of seeing and putting things. What ends up being labeled as criminal does not inhere in the act or the person. It is designed (or "consti-tuted" as Niels Christie, 2004, put it) through the act of interrogation:

> The world comes to us as we constitute it. Crime is thus a product of cultural, social and mental processes. For all acts, including those seen as unwanted, there are dozens of possible alternatives to their understanding: bad, mad, evil, misplaced honour, youth bravado, political heroism—or crime. The same acts can thus be met within several parallel systems as judicial, psychiatric, pedagogical, theological. (p. 10)

The same unwanted act (or "error"), in other words, can be construed to be many things at the same time, depending on which questions you asked at the beginning. Ask theological questions, and you may see in an error the manifestation of evil or the weakness of the flesh. Ask pedagogical questions, and you may see in the error the expression of underdeveloped skills. Ask judicial questions, and you may begin to see a crime. Unwanted acts do not contain something criminal as their essence. We make it so through the perspective we take, through the questions we ask.

A "Negligent" Surgeon in New Zealand

A British cardiothoracic surgeon, who had moved to New Zealand, was charged with manslaughter of three patients who had died during, or immediately after, operations that he had performed. A preceding inquiry had pointed to deficiencies in the surgeon's work. These cases were subsequently investigated by the police, which triggered criminal prosecution (Skegg, 1998).

Saying that the surgeon's acts amounted to incompetence, which in turn moti-vated criminal charges that converted those same acts into manslaughter, is one extreme way of dealing with medical failure. Other ways also are possible. For example, one could see this as an issue of cross-national transition (Are procedures for doctors moving to Australia or New Zealand from Europe adequate? And how are any cultural implications of practicing there systematically managed or monitored, if at all?). One could see it as a problem of access control to the profession (Do different countries have different standards for who they would want as a surgeon, and who controls access and how?); as one of training or proficiency checking (Do surgeons submit to regular and systematic follow-up of critical skills, such as the half-yearly proficiency check for professional pilots?); as an organizational one (the absence of regular junior staff to help with operations and to being obliged to work with medical students instead); or as sociopolitical (How is the assignment of resources and perhaps even oversight governed in facilities outside the capital?).

It may well be possible to write a compelling argument for each of these explanations of medical failure—each with a different repertoire of interpretations and countermeasures after it. Access and proficiency issues are controlled, out of the argument. Training problems are educated away from the issue. Organizational issues are managed away from the issue. Political problems are elected away from the issue. A crime is punished away from the issue. The point is not that one interpretation is right and all the others wrong. The point is that multiple overlapping interpretations of the same act are always possible, and that they have different ramifications for what individuals and organizations should do so that act does not happen or lead to bad consequences again.

The notion that crime is just one construction of an act, of many possible ones, is perhaps not easy to accept. We would think that a crime, of all things, must make up some essence behind a number of possible descriptions of an act, especially if that act has a bad outcome. We seem to have great confidence that the various descriptions can be sorted out by the rational process of a trial, that it will expose as patently false Christie's (2004) "psychiatric, pedagogical, theological" or organizational explanations ("I had failure anxiety!" "I wasn't trained enough!" "It was the Lord's will!" "I had lousy assistance, bad light, lack of sleep!"). Like a scalpel, the application of reason will strip away the noise, the decoys, and the excuses and arrive at the essential story: whether an act was negligent. And if negligence turns out not to make up the essence, then there will be no bad consequences. It should be that simple.

It is not. When we find an essence behind the complexity of an unwanted act with a bad outcome, it is not because that essence is there—independent, stable, and waiting for us to cut down to it—but because we created it as a result of the questions we asked and because we stopped looking any further once our construction was complete and fulfilled the social or political purposes we had in mind for it. As Christie (2004) argued, negligence, or crime, is not an essence that we can discover behind the inconsistency and shifting nature of the world as it meets us. Crime or negligence itself is that flux, that dynamism, that inconstancy, a negotiated arrangement, a tenuous and temporary stability achieved among shifting cultural, social, mental, and political forces. Concluding that an unwanted act is a crime is not the outcome of high-acuity observation. It is an accomplished human project, a social achievement.

Merton (1938) explored how social groups couple their desired ends (e.g., not having an accident happen, achieving safe performance) to moral and institutional regulation of permissible and required behavior (Morrill, Snyderman, & Dawson, 1997). Where the lines go between what is acceptable and what is not is constantly renegotiated at the intersection of societal, political, and technological (e.g., industrialization, urbanization, computerization) developments, giving different expressions to legality and illegality (Dekker, 2009; Foucault, 1977).

Sociological research into deviance (Goode, 1994; Rock, 1998) is thus more interested in those who draw the lines between acceptable and unacceptable behavior than those who cross them (Becker, 1963). Culpability, from this starting point, arises in part from people's ways of seeing and describing acts, something that not only evolves historically but also is situationally contingent (Christie, 2004). It has encouraged research into where the lines come from (Rafter, 1990), which can be seen in the work of Erikson (1966) and Foucault, who explicitly forced poststructuralist theory into criminal justice history with *Discipline and Punish* (Foucault, 1977). Who become moral entrepreneurs, imposing lines that separate legality from illegality, and how do these preserve or upset the status quo (Garland, 1993, 2002)? This is always an arena for political contest. It has made possible the idea of "overcriminalization" (Husak, 2008), something that people in safety-critical fields like healthcare would argue is happening now (International Civil Aviation Organization, 2007; Institute for Safe Medication Practices [ISMP], 2007).

A nurse from Wisconsin was charged with criminal "neglect of a patient causing great bodily harm" in the medication error-related death of a 16-year-old woman during labor. The nurse accidentally administered a bag of epidural analgesia by the intravenous route instead of the intended penicillin. The improbable moral entrepreneur in this case was an investigator for the Wisconsin Department of Justice, Medicaid Fraud Control Unit. The job of the fraud unit is to investigate and prosecute criminal offenses affecting the medical assistance program. This includes anything that affects the health, safety, and welfare of recipients of medical assistance. The teenage patient fell into that category. And with a stretch of the investigator's morally indignant imagination, the nurse's actions did as well (Dekker, 2010).

The constructionist lens to view the possible causes behind the increasing criminalization of professional mistakes in healthcare can be instructive (Engbersen & Van der Leun, 2001; Rafter, 1990), without necessarily defending that position other than as an analytical aid. It identifies possible social causes and psychosocial consequences. Healthcare should probably be seen as different from an area like road traffic (Tingvall & Lie, 2010), in which there is societal and political support for broad categories of negligence and recklessness, in part because of near-universal participation in the system and the large autonomy of individual actors in it (Amalberti, 2001). In a setting like road traffic, the Durkheimian function of criminalization (setting boundaries and demonstrating clearly to others where they go, *pour encourager les autres*, or "to encourage the others") is widely seen as meaningful (Erikson, 1966). The negative consequences of criminalization for safety, particularly detrimental effects on honest disclosure (Berlinger, 2005), also seem more articulated in healthcare than in these settings.

The social-constructionist argument does not explain specific shifts in societal assessments of criminality at specific times in history—only that such shifts occur, and that they, in general social terms, are the result of societal renegotiations in what is seen as sanctionable behavior. Why professionals in healthcare are perhaps more likely to be criminally prosecuted today as compared to, say, 40 years ago is not in itself explained.

The idea of an "accident" (and the concomitant growth of safety science and risk management) is relatively modern (Beck, 1992; Green, 2003). Until the scientific revolution in the seventeenth century, societies had little need for a concept like accident. Religion and superstition supplied explanatory models for misfortune, and where misfortune was going to occur was random, uncontrollable, unknowable. The notion that it was the result of divine or demonic incitement waned throughout the modern period, and was gradually replaced by a late nineteenth-century model that saw accidents as unfortunate but otherwise meaningless coincidences of space and time (Green, 2003). Over the last 30 years, however, the societal interpretation of accidents has shifted dramatically. Startling failures such as the collision of two jumbo jets at Tenerife in 1977 and the Three Mile Island nuclear accident in 1979 moved accidents back onto the center stage of our societies: Western society is said to be much more "risk conscious" (Wilkinson, 2001). Accidents are today seen as evidence that a particular risk was not managed well enough, and behind such mismanagement are people, individuals, or single acts of omission or commission (Bittle & Snider, 2006; Green, 2003).

The last 30 years has also seen a gradual reduction in the acceptance of risk altogether (Beck, 1992) and the expectation that some safety-critical activities are accident free, with a zero tolerance of failure. The increasingly flawless performance of some systems may have sponsored a societal belief in their infallibility and an intolerance of failure (Amalberti, 2001). Experts are expected to make any residual accidents comprehensible, which often means explaining which risk factors were not controlled and by whom. The accident has to go on somebody's account (Douglas, 1992). Note how societies have drifted from the idea of accident. Resources spent on formally investigating accidents would make no sense if accidents are truly accidental or random events.

Another feature of the last 30 years is the electronically mediated democratization and increasing accessibility of knowledge, as well as consumer vocalism and activism. These can put failings of complex systems like hospitals (or alleged failings of individuals in them) on fuller display than before ("Murder! Mayhem!" 2005). The media doubtlessly enjoy a strong role in celebrating certain accidents, while able to ignore others (Dekker, 2007b; Ditton & Duffy, 1983; Ödegård, 2007; Palmer, Emanuel, & Woods, 2001). A study linked cultural and political populism to the punitiveness of the criminal justice system of a country (Miyazawa, 2008), and media coverage of an event has been shown to articulate and animate social reactions to the point of constructing antiheroes (Elkin, 1955; McLean & Elkind, 2004) and their crimes (Dekker, 2007b; Ericson, 1995; Innes, 2004; Jacobs & Henry, 1996; Tuchman, 1978).

The coverage of and discourse surrounding social issues (e.g., hate crime, immigration, and by extension accidents and human error) have been linked to political populism, judicial responses, and the criminalization of new categories of human action (Blackwelder, 1996; Engbersen & Van der Leun, 2001; Husak, 2008; Jacobs & Henry, 1996; Phillips & Grattet, 2000). This could be seen as amounting to a strong democratic project (which defenders of media sensationalism in the wake of an accident or other undesirable social event likely would; "Murder! Mayhem!" 2005; Foucault, 1975), where the polity, through its judicial system, responds to and "fairly" represents the concerns of the society in which it operates.

As seems common in populist responses to perceived societal perils (Kieckhefer, 1976; Miyazawa, 2008), the constructed threat (e.g., medical error, hate crime) can be a stand-in for more diffuse social concerns (Becker, 1963; Ben-Yehuda, 1983; Foucault, 1975). Anxiety, or undifferentiated and undirected fear, is projected onto easily identifiable symbols of normative transgression: witches, gays, immigrants, terrorists, or any other "outsiders" (Becker, 1963).

Sociology has linked modern society and its anonymity and manifold uncertainties with anxiety—as a response to social processes and cultural experiences (Wilkinson, 2001). Disembedding (the decreasing relevance of place or locality), moral fragmentation and secularization, and concomitant fears of anomie (a wholesale erosion of norms and rules and adherence to them) are cited as sources of social anxiety in the late modern age (Giddens, 1991). According to this notion, expressing societal intolerance with pilot errors or drug misadministrations is related to such anxiety. Enhancing the visibility of such deviance by criminalizing it performs ancillary cultural work by highlighting moral boundaries (Rock, 1998), assuaging society's members that lines still exist or *should* still exist (Erikson, 1966; Foucault, 1975), consistent with links between populism, criminalization, and media sensationalism ("Murder! Mayhem!" 2005; Ditton & Duffy, 1983; Miyazawa, 2008).

CRIMINALIZING PROFESSIONAL ERROR IN HEALTHCARE: WHY A CONCERN?

One obvious concern about judicial action in the aftermath of error in healthcare has focused on how it interferes with the safety investigation by a hospital and destroys the willingness of people to voluntarily report errors and violations (Berlinger, 2005; Brous, 2008; Chapman, 2009; Dekker, 2007a, 2009; Flight Safety Foundation, 2006; Thomas, 2007). This is known to be a critical ingredient to the creation of "safety cultures": organizational cultures that encourage honest disclosure and open reflection on their own practices with the aim to constantly improve quality and safety of their products or services (Lauber, 1993). Such reflection, and the learning from failure that it can engender, is hampered when a professional mistake is criminalized:

> While there is considerable pressure from the public and the legal system to blame and punish individuals who make fatal errors, filing criminal charges against a healthcare provider who is involved in a medication error is unquestionably egregious and may only serve to drive the reporting of errors underground. The belief that a medication error could lead to felony charges, steep fines, and a jail sentence can also have a chilling effect on the recruitment and retention of healthcare providers—particularly nurses, who are already in short supply. (ISMP, 2007)

A common response to the threat of criminalization enacted spontaneously by professionals is to become better at making the evidence of mistake go away and not reporting errors: "practising under the threat of prosecution can only serve to hide errors" (Chapman, 2009). Another effect is the practice of "defensive medicine," which increases the use of unnecessary tests and procedures and fuels the rise in healthcare costs (Sharpe, 2004). Jointly, these effects create an adversarial stance that reduces openness and could be counterproductive to longer-term societal efforts to achieve a balance between learning and accountability in safety-critical systems.

Another important question is whether medical error can be sanctioned to disappear. Professional mistakes in healthcare are highly particular and contingent—anchored to and embedded in normal contexts in which people perform skilled work under conditions of resource constraints and outcome uncertainty. From this point of view, professional mistakes in healthcare can hardly be punished or sanctioned away—they are an inevitable part of the complex system in which they are generated (Vaughan, 1999). Errors and other undesired outcomes are the inevitable product of the structural interactive complexity and tight coupling of most safety-critical systems (Perrow, 1984); they emerge nonrandomly as antieffects from well-organized processes (Pidgeon & O'Leary, 2000) and might well be inevitable (Vaughan, 1996, 2005). For example, as was pointed out about drug errors: "Dispensing mistakes happen. And even with the introduction of robots and Standard Operating Procedures, the Utopian ideal of a world without errors is closer to fantasy than reality" (Chapman, 2009).

Finally, the prosecution of professionals can distort the allocation of scarce societal resources within the criminal justice system (Jacobs & Henry, 1996) when there are already bodies in place (e.g., peer groups, medical discipline committees) that could be better positioned to deal effectively with the aftermath of failure in healthcare. In addition, systemic interventions (through new technology) are commonly known to have better safety effects than the prosecution of individuals. This is likely true notwithstanding the possible perils of technology as discussed in Chapter 4. For example:

> The addition of anti-hypoxic devices to anesthetic machines and the widespread adoption of pulse oximetry have been much more effective in reducing accidents in relation to the administration of adequate concentrations of oxygen to anesthetized patients than has the conviction for manslaughter of an anesthetist who omitted to give oxygen to a child in 1982. (Merry & Peck, 1995)

Naturally, victims may derive some measure of solace, if not a sense of retribution, with the criminalization of professional mistakes. Yet criminalization of an individual can also be seen by victims as unfair and counterproductive or as scapegoating (Mellema, 2000). Even victims might interpret this as getting the organization or government regulators off the hook and oversimplifying the complexity of contributory events. This is also discussed in the safety literature (Perrow, 1984) and literatures on healthcare (Beaver, 2002; Osborne, Blais, & Hayes, 1999), in which condensed explanations of failure and concomitant criminalization are used to protect elite interests (Levack, 1987) and avoid the costs of fixing or retrofitting a system (Goode, 1994). In addition, criminalizing an individual may not give victims the confidence that a similar incident will be prevented in the future (Dekker, 2007d; Dekker & Hugh, 2009; Merry & McCall Smith, 2001). The mother of a 3-month-old killed as a result of medication misadministration, for instance, stopped seeing the point of the criminal trial against the nurse long before the proceedings had concluded in a guilty verdict (Ödegård, 2007).

THE SECOND VICTIM

> Virtually every practitioner knows the sickening realization of making a bad mistake. You feel singled out and exposed—seized by the instinct to see if anyone has noticed. You agonize about what to do, whether to tell anyone, what to say. Later, the event replays itself over and over in your mind. You question your competence but fear being discovered. You know you should confess, but dread the prospect of potential punishment and of the patient's anger. You may become overly attentive to the patient or family, lamenting the failure to do so earlier and if you haven't told them, wondering if they know. (Wu, 2000, p. 726)

For healthcare professionals, an error that leads to an incident or death is antithetical to their identities, a devastating failure to live up to their deontological commitment (Wolf, 1994). The memory of error stays with professionals for many years (Serembus, Wolf, & Youngblood, 2001). All of these effects are visible and can be strongly expressed *before* any organizational sanction, civil suit, or criminal prosecution. It could be argued that people punish themselves harshly in the wake of failure, and that society or organizations cannot make that much worse. Having made an error in the execution of a job that involves error management and prevention is something that causes excessive stress, depression, anxiety, and other psychological ill-health (Berlinger, 2005).

Particularly when the work involves considerable autonomy and presumptions of control over outcomes on the part of the actor, guilt and self-blame are common, with professionals often denying the role of the system or organization in the spawning of their error altogether and blaming themselves entirely (Meurier, Vincent, & Parmar, 1998; Snook, 2000). This sometimes includes hiding the error or its consequences from family and friends, the professionals distancing themselves from any possible support, and attempting to make atonement on one's own accord with those harmed by the error (Christensen, Levinson, & Dunn, 1992).

Criminalization of medical error and other forms of official sanction (medical responsibility boards, civil lawsuits) affirm feelings of guilt and self-blame and exacerbate their effects, which are the sorts of effects that are linked to poor clinical outcomes in other criminological settings (Friel, White, & Alistair, 2008). In the case of criminalizing human error, it can lead to people departing on sick leave, divorcing, exiting the profession permanently, or committing suicide (Meszaros & Fischer-Danzinger, 2000; Tyler, 2003). Another response, although much more rare, is an expression of anger and counterattack, for example, through the filing of a defamation lawsuit (Anderson, 2005; Sharpe, 2004).

Criminalization can also have consequences for a person's livelihood (and his or her family), as licenses to practice may be revoked automatically (although, perversely, not always; Ödegård, 2007), which in turn can generate a whole new layer of anxiety and stress. One pharmacist, whose medication error ended in the death of two patients, suffered from depression and anxiety to such an extent that he eventually stabbed his wife to death and injured his daughter with a knife (Serembus et al., 2001).

In the best case, professionals seek to process and learn from the mistake, discussing details of their error with their peers or employer, contributing to its systematic investigation and helping with putting safety checks in place (Christensen et al., 1992). The role of the organization in facilitating such coping (e.g., through peer and managerial support and appropriate structures and processes for learning from failure) is hugely important, as was demonstrated, for example, in a longitudinal study in a large safety-critical facility (Dekker & Laursen, 2007).

PEER AND EMPLOYEE ASSISTANCE

Research on employee assistance programs has suggested that it is crucial that employees do not get constructed as if they are the source of the problem and treated as somehow "troubled" as opposed to "normal" employees (Cooper & Payne, 1988; Dekker & Laursen, 2007). If this condition is met, employee support, particularly peer support, appears to be one of the most important mediating variables in managing stress, anxiety, and depression in the aftermath of error and one of the strongest predictors of coming out psychologically healthy (Dekker & Laursen, 2007).

Peer support or employee assistance programs are not yet widespread in healthcare, however. As Wu (2000) said:

> Sadly, the kind of unconditional sympathy and support that are really needed are rarely forthcoming. While there is a norm of not criticizing, reassurance from colleagues is often grudging or qualified. ... In the absence of mechanisms for healing, physicians find dysfunctional ways to protect themselves. They often respond to their own mistakes with anger and projection of blame, and may act defensively or callously and blame or scold the patient or other members of the healthcare team. (pp. 726–727)

Other fields have learned that for a peer program to be effective, it should be made clear that the peer support program is run and administered by peers and explicitly not by the employer. Experience suggests that employer involvement makes the affected worker quickly doubt the purity of the motives for offering help (Leonhardt & Vogt, 2006). Employers have liability to manage (e.g., this is often part of the role of the risk manager of the hospital in the wake of medical failure). It has a reputation to uphold, politicians to keep at bay, and production targets to make. This could make employees feel pressured into one or another kind of settlement or agreement. They may fear that their employment can be terminated or that any support is simply part of a cynical endeavor to get them back into the saddle more quickly than they are capable of handling or than is good for their patients. Criminalization, of course, frustrates the possibility to intervene on the part of either employers or peers, particularly when the professional is incarcerated.

Such work on peer assistance programs suggests that much can be done inside an organization to foster a climate of openness and learning (and indeed, a perception of justice, trust, and care) that can be independent of what happens in the judicial sphere of that organization. Whereas the judicial climate in a country can discourage open reporting and honest disclosure, this does not mean that a healthcare organization cannot try to build a basis for a just culture. A whole host of practitioner needs becomes active in the wake of involvement in an adverse event, and keeping out of trouble may not even be first on everybody's mind. Attending to practitioner needs is an obvious way in which an organization can show that it cares.

To be sure, practitioners can themselves be their own worst enemies when it comes to dealing with the consequences of an adverse event. Professional and cultural dispositions may conspire against their ability to put things in context and perspective. The expectation of perfection clashes both psychologically and practically with any efforts to learn from mistakes. Leape (1994) called this a paradox:

> The paradox is that although the standard of medical practice is perfection—error-free patient care—all physicians recognize that mistakes are inevitable. Most would like to examine their mistakes and learn from them. From an emotional standpoint, they need the support and understanding of their colleagues and patients when they make mistakes. Yet they are denied both insight and support by misguided concepts of infallibility and by fear: fear of embarrassment by colleagues, fear of patient reaction, and fear of litigation. Although the notion of infallibility fails the reality test, the fears are well grounded. (p. 1852)

This means that hospitals or other healthcare organizations have all kinds of obstacles to overcome in their effort to increase learning from adverse events. And all these efforts are set against a background of people's possible uncertainty about how the organization is going to respond when it learns about an adverse event. Removing, as much as possible, uncertainty about who gets to draw the line (and resolving perceived unfairness of who gets to do that) is critical.

KEY POINTS

- Attributing adverse events to human error does little to improve the healthcare system. Blame inhibits learning. Simple attributions of an adverse event to human error stop deeper investigation and hamper understanding. But not all errors are seen as equal, or equally blameworthy.
- To know which errors deserve which response, the primitive belief in healthcare is that we just need to distinguish among human error, at-risk behavior, and reckless behavior. But those categories are nothing more than our own judgments, our own attributions, which are forever contestable by those who judge things differently.
- Assigning acts to categories is a matter of power. Who has the power to say that behavior is one thing and not the other? And who has the power to decide on the response? As soon as power is involved in categorizations, responses can be seen as unjust, as unfair. Justice, after all, is a perception.

The way of getting to the response (e.g., who gets to participate, who is heard and how) also can be seen as just or unjust and often forms an important leverage point for the just culture of an organization.

- Blame-free or no-fault organizations are not accountability free. Holding people accountable and blaming them are two different things. Blaming people may in fact make them less accountable. They might tell fewer accounts; they may feel less compelled to have their voice heard or to participate in improvement efforts. And they may feel more motivated to hide the evidence of errors.

- For healthcare professionals, an error that leads to an incident or death is antithetical to their identities. It can cause excessive stress, depression, anxiety, and other psychological ill-health. With work that involves considerable autonomy and control over outcomes, guilt and self-blame are common. Practitioners often deny the role of the system or organization in the spawning of their error. Criminalization of error (with the judicial system getting involved) exacerbates all these undesirable effects.

REFERENCES

Alicke, M. D. (2000). Culpable control and the psychology of blame. *Psychological Bulletin, 126*(4), 556–556.

Amalberti, R. (2001). The paradoxes of almost totally safe transportation systems. *Safety Science, 37*(2–3), 109–126.

Anderson, R. E. (2005). *Medical malpractice: A physician's sourcebook.* New York: Humana Press.

Beaver, E. (2002). Witchcraft, female aggression, and power in the Early Modern community. *Journal of Social History, 35*(4), 955–988.

Beck, U. (1992). *Risk society: Towards a new modernity.* London: Sage.

Becker, H. S. (1963). *Outsiders: Studies in the sociology of deviance.* London: Free Press of Glencoe.

Ben-Yehuda, N. (1983). The European witch craze: Still a sociologist's perspective. *American Journal of Sociology, 88*(6), 1275–1279.

Berlinger, N. (2005). *After harm: Medical error and the ethics of forgiveness.* Baltimore: Johns Hopkins University Press.

Bittle, S., & Snider, L. (2006). From manslaughter to preventable accident: Shaping corporate criminal liability. *Law and Policy, 28*(4).

Bjarup, J. (2005). The philosophy of Scandinavian legal realism. *Ratio Juris, 18*(1), 1–15.

Blackwelder, S. P. (1996). Fearless wives and frightened shrews: The construction of the witch in Early Modern Germany. *Contemporary Sociology, 25*(4), 525–563.

Bosk, C. (2003). *Forgive and remember: Managing medical failure.* Chicago: University of Chicago Press.

Brous, E. (2008). Criminalization of unintentional error: Implications for TAANA. *Journal of Nursing Law, 12*(1), 5–12.

Chapman, C. (2009, April 18). A criminal mistake? *Chemist and Druggist,* 8.

Christensen, J. F., Levinson, W., & Dunn, P. M. (1992). The heart of darkness: The impact of perceived mistakes on physicians. *Journal of General Internal Medicine, 7,* 424–431.

Christie, N. (2004). *A suitable amount of crime.* London: Routledge.

Cook, R. I., & Nemeth, C. P. (2010). "Those found responsible have been sacked": Some observations on the usefulness of error. *Cognition, Technology and Work, 12,* 87–93.

Cooper, C. L., & Payne, R. (1988). *Causes, coping, and consequences of stress at work.* Chichester, UK: Wiley.

Dekker, S. W. A. (2002). *The field guide to human error investigations.* Bedford, UK: Cranfield University Press.

Dekker, S. W. A. (2006). *The field guide to understanding human error.* Aldershot, UK: Ashgate.

Dekker, S. W. A. (2007a). Criminalization of medical error: Who draws the line? *ANZ Journal of Surgery, 77*(10), 831–837.

Dekker, S. W. A. (2007b). Discontinuity and disaster: Gaps and the negotiation of culpability in medication delivery. *Journal of Law, Medicine and Ethics, 35*(3), 463–470.

Dekker, S. W. A. (2007c). Doctors are more dangerous than gun owners: A rejoinder to error counting. *Human Factors, 49*(2), 177–184.

Dekker, S. W. A. (2007d). *Just culture: Balancing safety and accountability.* Aldershot, UK: Ashgate.

Dekker, S. W. A. (2009). Just culture: Who draws the line? *Cognition, Technology and Work, 11*(3), 177–185.

Dekker, S. W. A. (2010). We have Newton on a retainer: Reductionism when we need systems thinking. *The Joint Commission Journal on Quality and Patient Safety, 36*(4), 147–149.

Dekker, S. W. A., & Hugh, T. B. (2009). Balancing "no blame" with accountability in patient safety. *New England Journal of Medicine, 362*(3), 275.

Dekker, S. W. A., & Laursen, T. (2007). From punitive action to confidential reporting: A longitudinal study of organizational learning. *Patient Safety and Quality Healthcare, 5,* 50–56.

Ditton, J., & Duffy, J. (1983). Bias in the newspaper reporting of crime news. *British Journal of Criminology, 23*(2), 159–165.

Douglas, M. (1992). *Risk and blame: Essays in cultural theory.* London: Routledge.

Elkin, F. (1955). Hero symbols and audience gratifications. *Journal of Educational Sociology, 29*(3), 97–107.

Engbersen, G., & Van der Leun, J. (2001). The social construction of illegality and criminality. *European Journal on Criminal Policy and Research, 9,* 51–70.

Ericson, R. (Ed.). (1995). *Crime and the media.* Dartmouth, UK: Aldershot.

Erikson, K. (1966). *Wayward Puritans: A study in the sociology of deviance.* New York: Wiley.

Flight Safety Foundation. (2006). *Aviation safety groups issue joint resolution condemning criminalization of accident investigations.* Washington, DC: Author.

Foucault, M. (1975). *The spectacle of the scaffold.* London: Penguin Group.

Foucault, M. (1977). *Discipline and punish: The birth of the prison.* New York: Pantheon Books.

Friel, A., White, T., & Alistair, H. (2008). Posttraumatic stress disorder and criminal responsibility. *Journal of Forensic Psychiatry and Psychology, 19*(1), 64–85.

GAIN (2004). Roadmap to a just culture: Enhancing the safety environment. Global Aviation Information Network (Group E: Flight Ops/ATC Ops Safety Information Sharing Working Group), 3

Garland, D. (1993). *Punishment and modern society: A study in social theory.* Chicago: University of Chicago Press.

Garland, D. (2002). *The culture of control: Crime and social order in contemporary society.* Chicago: University of Chicago Press.

Giddens, A. (1991). *Modernity and self-identity: Self and society in the late modern age.* Stanford, CA: Stanford University Press.

Goode, E. (1994). Round up the usual suspects: Crime, deviance, and the limits of constructionism. *American Sociologist, 25,* 90–104.

Green, J. (2003). The ultimate challenge for risk technologies: Controlling the accidental. In J. Summerton & B. Berner (Eds.), *Constructing risk and safety in technological practice.* London: Routledge.

Hugh, T. B., & Dekker, S. W. A. (2009). Hindsight bias and outcome bias in the social construction of medical negligence: A review. *Journal of Law and Medicine, 16*(5), 846–857.

Husak, D. (2008). *Overcriminalization: The limits of the criminal law*. New York: Oxford University Press.

Innes, M. (2004). Crime as a signal, crime as a memory. *Journal for Crime, Conflict and the Media, 1*(2), 15–22.

Institute of Medicine. (2003). *Patient safety: Achieving a new standard for care*. Washington, DC: National Academy of Sciences, Institute of Medicine.

Institute for Safe Medication Practices. (2007). *Criminal prosecution of human error will likely have dangerous long-term consequences*. Retrieved February 4, 2010, from http://www.ismp.org/Newsletters/acutecare/articles/20070308.asp

International Civil Aviation Organization. (2007). *Working paper of the 36th session of the Assembly: Protection of certain accident and incident records and of safety data collection and processing systems in order to improve aviation safety* (Vol. A36-WP/71 TE/16, 17/8/07). Montreal: Author.

Jacobs, J. B., & Henry, J. S. (1996). The social construction of a hate crime epidemic. *The Journal of Criminal Law and Criminology, 86*(2), 366–391.

Kieckhefer, R. (1976). *European witch trials, their foundations in popular and learned culture*. London: Routledge.

Lauber, J. K. (1993). *A safety culture perspective*. Paper presented at the Flight Safety Foundation's 38th Corporate Aviation Safety Seminar, April 1993, Irving, Texas.

Leape, L. L. (1994). Error in medicine. *Journal of the American Medical Association, 272*(23), 1851–1857.

Leonhardt, J., & Vogt, J. (2006). *Critical incident stress management in aviation*. Aldershot, UK: Ashgate.

Lerner, J. S., & Tetlock, P. E. (1999). Accounting for the effects of accountability. *Psychological Bulletin, 125*(2), 255–275.

Levack, B. P. (1987). *The witch-hunt in early modern Europe*. London: Longman.

Mahler, J. G. (2009). *Organizational learning at NASA: The* Challenger *and* Columbia *accidents*. Washington, DC: Georgetown University Press.

Marx, D. (2001). *Patient safety and the "just culture": A primer for health care executives*. New York: Columbia University Press.

McLean, B., & Elkind, P. (2004). *The smartest guys in the room: The amazing rise and scandalous fall of Enron*. New York: Portfolio.

Mee, C. L. (2007). Should human error be a crime? *Nursing, 37*, 6.

Mellema, G. (2000). Scapegoats. *Criminal Justice Ethics*, 3–9.

Merry, A. F., & McCall Smith, A. (2001). *Errors, medicine and the law*. Cambridge, UK: Cambridge University Press.

Merry, A. F., & Peck, D. J. (1995). Anaesthetists, errors in drug administration and the law. *New Zealand Medical Journal, 108*, 185–187.

Merton, R. K. (1938). Social structure and anomie. *American Sociological Review, 3*(5), 672–682.

Meszaros, K., & Fischer-Danzinger, D. (2000). Extended suicide attempt: Psychopathology, personality and risk factors. *Psychopathology, 33*(1), 5–10.

Meurier, C. E., Vincent, C. A., & Parmar, D. G. (1998). Nurses' responses to severity dependent errors: A study of the causal attributions made by nurses following an error. *Journal of Advanced Nursing, 27*, 349–354.

Miyazawa, S. (2008). The politics of increasing punitiveness and the rising populism in Japanese criminal justice policy. *Punishment and Society, 10*(1), 47–77.

Moran, D. (2008). I was treated like a criminal after a harmless drug error. *Nursing Standard, 22*, 33.

Morath, J. M., & Turnbull, J. E. (2005). *To do no harm: Ensuring patient safety in health care organizations.* San Francisco: Jossey-Bass.

Morreim, E. H. (2004). Medical errors: Pinning the blame versus blaming the system. In V. A. Sharpe (Ed.), *Accountability: Patient safety and policy reform* (pp. 213–232). Washington DC: Georgetown University Press.

Morrill, C., Snyderman, E., & Dawson, E. J. (1997). It's not what you do, but who you are: Informal social control, social status, and normative seriousness in organizations. *Sociological Forum, 12*(4), 519–543.

Murder! Mayhem! Social order! (2005). *Wilson Quarterly, 29,* 94–96.

North, D. M. (2000). Let judicial system run its course in crash cases. *Aviation Week and Space Technology, 152*(20), 66–67.

Ödegård, S. (Ed.). (2007). *I rättvisans namn* [In the name of justice]. Stockholm: Liber.

Osborne, J., Blais, K., & Hayes, J. S. (1999). Nurses' perceptions: When is it a medication error? *Journal of Nursing Administration, 29*(4), 33–38.

Palmer, L. I., Emanuel, L. L., & Woods, D. D. (2001). *Managing systems of accountability for patient safety.* Washington, DC: National Health Care Safety Council of the National Patient Safety Foundation.

Pandit, M. S. (2009). Medical negligence: Criminal prosecution of medical professionals, importance of medical evidence: Some guidelines for medical practitioners. *Indian Journal of Urology, 25*(3), 379–383.

Pellegrino, E. D. (2004). Prevention of medical error: Where professional and organizational ethics meet. In V. A. Sharpe (Ed.), *Accountability: Patient safety and policy reform* (pp. 83–98). Washington, DC: Georgetown University Press.

Perrow, C. (1984). *Normal accidents: Living with high-risk technologies.* New York: Basic Books.

Phillips, S., & Grattet, R. (2000). Judicial rhetoric, meaning-making, and the institutionalization of hate crime law. *Law and Society Review, 34*(3), 567–606.

Pidgeon, N., & O'Leary, M. (2000). Man-made disasters: Why technology and organizations (sometimes) fail. *Safety Science, 34*(1–3), 15–30.

Rafter, N. H. (1990). The social construction of crime and crime control. *Journal of Research in Crime and Delinquency, 27,* 376–389.

Reason, J. T. (1997). *Managing the risks of organizational accidents.* Farnham, UK: Ashgate.

Reissner, D. (2009). A criminal mistake? *Chemist and Druggist, 271*(6693), 8–9.

Rock, P. (1998). Rules, boundaries and the courts: Some problems in the Neo-Durkheimian sociology of deviance. *The British Journal of Sociology, 49*(4), 586–601.

Serembus, J. F., Wolf, Z. R., & Youngblood, N. (2001). Consequences of fatal medication errors for healthcare providers: A secondary analysis study. *MedSurg Nursing, 10*(4), 193–201.

Sharpe, V. A. (2003). Promoting patient safety: An ethical basis for policy deliberation. *Hastings Center Report, 33*(5), S2–S19.

Sharpe, V. A. (2004). *Accountability: Patient safety and policy reform.* Washington, DC: Georgetown University Press.

Skegg, P. D. G. (1998). Criminal prosecutions of negligent health professionals: The New Zealand experience. *Medical Law Review, 6*(2), 220–246.

Snook, S. A. (2000). *Friendly fire: The accidental shootdown of U.S. Black Hawks over northern Iraq.* Princeton, NJ: Princeton University Press.

Summerton, J., & Berner, B. (2003). *Constructing risk and safety in technological practice.* London: Routledge.

Thomas, G. (2007). A crime against safety. *Air Transport World, 44,* 57–59.

Tingvall, C., & Lie, A. (2010). *The concept of responsibility in road traffic* [Ansvarsbegreppet i vägtrafiken]. Paper presented at the Transportforum, January 2010, Linkoeping, Sweden.

Tuchman, G. (1978). *Making news: A study in the construction of reality.* New York: Free Press.

Tyler, K. (2003). Helping employees cope with grief. *HRMagazine, 48*(9), 54–58.

Vaughan, D. (1996). *The* Challenger *launch decision: Risky technology, culture, and deviance at NASA*. Chicago: University of Chicago Press.

Vaughan, D. (1999). The dark side of organizations: Mistake, misconduct, and disaster. *Annual Review of Sociology, 25*, 271–305.

Vaughan, D. (2005). System effects: On slippery slopes, repeating negative patterns, and learning from mistake? In W. H. Starbuck & M. Farjoun (Eds.), *Organization at the limit: Lessons from the* Columbia *disaster* (pp. 41–59). Malden, MA: Blackwell.

Weiner, B. J., Hobgood, C., & Lewis, M. A. (2008). The meaning of justice in safety incident reporting. *Social Science and Medicine, 66*, 403–413.

Wilkinson, I. (2001). *Anxiety in a risk society*. New York: Routledge.

Wolf, Z. R. (1994). *Medication errors: The nursing experience*. Albany, NY: Delmar.

Woods, D. D., Dekker, S. W. A., Cook, R. I., Johannesen, L. J., & Sarter, N. B. (2010). *Behind human error*. Aldershot, UK: Ashgate.

Wu, A. W. (2000). Medical error: The second victim. *British Medical Journal, 320*(7237), 726–728.

8 New Frontiers in Patient Safety
Complexity and Systems Thinking

Most people in healthcare readily embrace the idea that they work in a complex system. The concept of healthcare as a complex system seems to be widely recognized (Pisek & Greenhalgh, 2001). But do we know what we mean when we say *complex*? How is complex different from complicated, for example? And for that matter, do practitioners and managers in healthcare typically act in ways that are appropriate for complex systems? There is a strong tendency to reach for simple solutions, for silver bullets, for single-factor explanations. And it is still common to bemoan the "ineptitude" of those defeated by the complexity of the system (Gawande, 2002) or to celebrate the "strength of character" of those able to bring it under control (Pellegrino, 2004). There is a lingering focus, in other words, on good and bad components rather than on the system. Of course, this may be inextricably connected to issues of identity and competence in healthcare, as discussed in Chapter 1. But focusing on components is something that can work in simple, or merely complicated, systems. It is useless in complex systems.

So, what is complexity, and how does it represent a completely new angle on our understanding of healthcare and patient safety? There are different readings of complexity that stem from different traditions and epistemologies (e.g., computational, social), but all attempt to describe the nature of complex systems and agree on a number of things. Complex systems consist of numerous components or agents that are interrelated in all kinds of ways, and they are open systems. They keep changing in interaction with their environment, and their boundaries are fuzzy. It can be hard to find out (or it is ultimately arbitrary) where the system ends and the environment begins. More than one description of complex systems is always possible and even necessary, even though the system will probably have changed before any description is finished.

There is no intelligent designer or governor who put a complex system together or controls it—complexity arises because of local interactions. In fact, complex systems are held together by local interactions only. The horizon of each component is limited, and the further away it is, the more unpredictable the consequences of its actions become. If there were one component that understood the whole system, then that component would have to be as complex as the complex system, which is a practical and philosophical impossibility (or possible only if the system were not complex). In a complex system, because of the deep and extended webs of inter-actions and interconnections, the action of any agent controls little but influences almost everything.

Complexity theory does not necessarily provide the exact tools with which to solve complex problems (in fact, that sheer possibility is antithetical to complex-ity), but it can provide rigorous accounts of why complex problems are so difficult (Cilliers, 2005). As Cilliers put it:

> Because complex systems are open systems, we need to understand the system's com-plete environment before we can understand the system, and of course, the environ-ment is complex in itself. There is no human way of doing this. The knowledge we have of complex systems is based on the models we make of these systems, but in order to function as models—and not merely as a repetition of the system—they have to reduce the complexity of the system. This means that some aspects of the system are always left out of consideration. The problem is confounded by the fact that that which is left out, interacts with the rest of the system in a non-linear way and we can therefore not predict what the effects of our reduction of the complexity will be, especially not as the system and its environment develops and transforms in time. (p. 258)

COMPLICATED VERSUS COMPLEX

The differences between complicated and complex are instructive. They can be summed up as follows: Complicated systems are ultimately knowable. They afford a complete, exhaustive description. A set of equations can be drawn up that fully captures their workings. Because of this, complicated systems are controllable, like machines are. Order in complicated systems is achieved by figuring out one best (e.g., efficient) method to operate them. Stability is achieved by compliance with this one best method. A jet airliner is a complicated system. Complex systems, in contrast, are never fully knowable. A complete, exhaustive description is impossible to attain, and these systems are mathematically intractable. No set of equations can ever capture their nature or full workings. Order is not imposed; it emerges from the multitude of relationships and interactions between component parts. Success in a complex system flows not from having it follow one best method but from a diversity of responses that allow it to cope with a changing environment. Mayonnaise is a complex system. Once created, it is a unique creation of emergent chemical prop-erties. It cannot be created, deconstructed, and reconstituted as from its original constituent parts. When open to the environment, it will also change its properties. (Cilliers, 1998).

Recall the example from obstetric intervention decisions in Chapter 6. The intro-duction of so-called ST-wave analysis technology there was intended to help improve intervention decisions during labor. Like in other medical decision making, the role of new technology in obstetric intervention decisions can become much clearer by the distinction between complicated and complex. It leads to interesting conclusions about the limits of our epistemological reach in the pursuit of technological or mana-gerial improvements in healthcare.

OBSTETRICS, INTERVENTION DECISIONS, AND NEW TECHNOLOGY

For a variety of anatomical and physiological reasons, labor is still dangerous for both fetus and parturient, even in the West (Amer-Wahlin & Dekker, 2008). Hypoxic injury to the child is one important risk, and fetal monitoring aims to capture its early signs. Traditionally, fetal monitoring is done through cardiotocography (CTG) via an abdominal Doppler on the mother or a fetal scalp electrode on the baby. A historical-graphical representation of the baby's heart rate and uterine contractions, CTG is both highly sensitive (good at detecting true positives) and unspecific (it also generates many false positives). So, it can be complemented with fetal blood sampling to identify metabolic acidosis in the baby, reflected by a low fetal blood pH, which is in turn an indication of inadequate fetal oxygenation. Interventions in the labor process may be necessary on a variety of clinical indications, of which fetal hypoxia is an important one. Intervention can range from offering drugs to speed labor to removing the fetus by emergency cesarean section (Drife & Magowan, 2004).

Intervention decisions (when to increase inputs to, or take over from, an autonomous process, for example) are fundamental to many ergonomic problems (Kerstholt, Passenier, Houttuin, & Schuffel, 1996). They are also fundamentally problematic because, particularly in escalating situations, they come either too early or too late. An early intervention may delete the evidence of a problem that warranted one, whereas a late intervention may lag behind any meaningful ability to retain or reestablish process integrity (Dekker & Woods, 1999b). The difficulty in timing interventions right may be reflected, for instance, in the increase of cesarean sections in a number of U.S. states, despite postoperative risks to the mother and a lack of convincing clinical indications (Zaccaria, 2002).

Obstetric technology is now available that can analyze a fetal electrocardiogram (ECG) by dividing the QRS complex of the ECG through its T wave (the T wave represents repolarization of fetal cardiac ventricles; the QRS complex corresponds to their depolarization). A rise in T-wave amplitude relative to the QRS waveform warns of compromised fetal adaptation to hypoxia (Amer-Wahlin, Yli, & Rosen, 2009) and is an earlier sign of trouble than what can be obtained through fetal blood sampling. It is also thought to be more specific than the CTG curve.

Vendors purvey the technology through promises to make intervention decisions both easier and better, focus clinicians' attention on the right answer, avoid unnecessary cesarean sections or other difficult instrumental deliveries, and provide reassurance to clinicians when interventions are not necessary (Amer-Wahlin, Ingemarsson, Marsal, & Herbst, 2005). The disconnect between the promises of new technology on the one hand and its problems and real potential on the other (Wiener, 1988) maps onto the complicated-complex divide. Vendors assume the system is merely complicated when they promise better results through the replacement of human work with computerized aids (components can be substituted with predictable results), when they promise to offload humans (there is a fixed amount of knowable work the system needs to do; if the machine does more, the human does less), and when they say they can focus human attention (there is a right answer that they already know).

In the past two decades or more, however, new technology not only typically entered complex (rather than complicated) fields of practice, but also it created reverberations typical of complexity. New roles emerge and relationships between people and artifacts are transformed (Woods & Dekker, 2000). Interconnections between people, departments, things, and systems proliferate (Perry, Wears, & Cook, 2005). New kinds of human work are produced that call on new sorts of expertise. And as a result, new technology opens up new pathways to failure that can be hard to foresee (Cook, Potter, Woods, & McDonald, 1991; Cook & Woods, 1996a, 1996b; Dekker & Woods, 1999a; Norman, 1990; Sarter, Woods, & Billings, 1997).

SIGNAL DETECTION IN OBSTETRICS

Systems that monitor selected process data (like a CTG curve) for potentially abnormal conditions (e.g., indications of fetal hypoxia) can be modeled as signal detection devices (Sorkin & Woods, 1985). These aim to discriminate between noise and signal plus noise. Signal detector systems take measurements (e.g., fetal CTG and other indicators of patient condition) that together form a multidimensional input vector called X and then compute a unidimensional statistic called Z based on the input vector compared to stored information about thresholds that relate to the expected characteristics of noise versus signal-plus-noise events in the monitored process. In obstetrics, for example, multidimensional inputs of CTG quality, patient condition, and history are compared against various stated numeric safety boundaries for heart rate, heart rate variations, and fetal blood pH (these are often contained in clinical guidelines). Signal detection theory assumes the Z distributions for signal and noise to be Gaussian and of equal variance, which seems a close approximation of many detection systems (Sorkin & Woods, 1985) (Figure 8.1).

Signal detection theory separates the monitoring problem into two distinct variables: a sensitivity parameter d' and a response criterion β. *Sensitivity* refers to the ability of the system to discriminate signal from noise on the input channel X. d' is calculated as a function of the normalized distance between the means of the two Z distributions (see Equation 8.1):

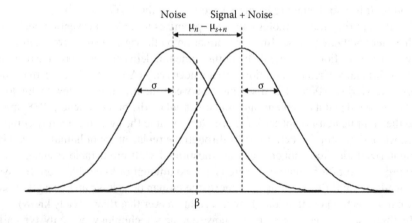

FIGURE 8.1   Noise and signal-plus-noise distributions of Z.

$$d' = (\mu n - \mu s + n)/\sigma \qquad (8.1)$$

In obstetrics, this sensitivity can be influenced by accuracy or resolution of a CTG readout that is consulted by the human or machine monitor, for example. The difference between a CTG generated by abdominal Doppler versus an FSE is a difference in sensitivity, or d'.

The Z statistic is compared against a response criterion β, which specifies how much evidence is required for the signal detector to decide on the presence of a signal (e.g., impaired fetal compensation for hypoxia). β can be moved as a result of payoffs (e.g., with high malpractice risk, β might be set low, demanding only some evidence of signal for intervention) and prior knowledge of probabilities (a product of training and experience). Unless the system is perfectly sensitive (i.e., there is no overlap between the noise and signal-plus-noise distributions), there is no optimal placement of β. With partially overlapping distributions, β is always a compromise between the probabilities of hits, correct rejections, misses, and false alarms. Its placement is a response to the costs and benefits as well as historical probabilities of each.

The insertion of a machine monitor (such as ST waveform technology) between human and monitored process in a sense doubles the signal detection problem (Sorkin & Woods, 1985). The automated subsystem (ST waveform technology) monitors a noisy input channel (fetal ECG) for occasional signal events (which are defined according to an algorithm that essentially compares the input vector of the machine X_a with prespecified thresholds). The automated monitor itself has a particular sensitivity (or d'_a) and a decision criterion β_a. If Z_a is equal to or greater than β_a, the machine will create an output (an ST event alarm) that is sent to the human monitor. This is represented by Figure 8.2. Predictably, if the probability of a machine output is high (e.g., through a high d'_a or low β_a), d'_h will decrease,

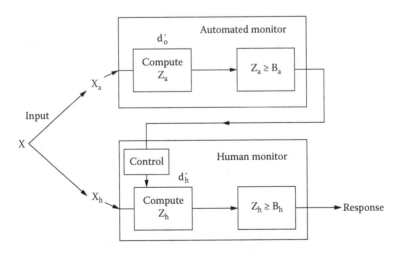

FIGURE 8.2 Automated and human monitor in sequence, yet both take their own input vector X from the monitored process. (Adapted from Sorkin, R. D., & Woods, D. D., Systems with human monitors: A signal detection analysis. *Human-Computer Interaction, 1*(1), 49–75, 1985.)

sometimes problematically referred to as "complacency" (Moray & Inagaki, 2000), whereas β_h will increase. Given that β_a is often set to minimize the number of missed signals (and can be manipulated without much cost or computing capacity, quite unlike d'_a), the overall performance of the monitoring ensemble is likely suboptimal (Sorkin & Woods, 1985).

COMPLEXITY AND SIGNAL DETECTION

The description in the preceding section is that of a complicated system. It can, in principle, be completed through an analysis of the behavior of its components (the machine and human monitors). Replacing a part of one (human judgment) by the other (ST waveform monitoring and alarm generation) creates better or different— but still predictable—results, for example, the cross interactions between the respective d' and β. Remarkably, this explained little of the obstetric intervention decisions we observed (Amer-Wåhlin, Bergström, Wahren, & Dekker, 2010). The socially noisy and cognitively intricate work of obstetrics seemed to quickly turn the merely complicated into something intractably complex.

X_a and X_h are virtually incomparable. Whereas X_a in ST waveform technology relies on a relatively coarse division of T through QRS in a fetal ECG, X_h (the expert human's input vector on which intervention decisions are based) constitutes an impossibly rich and constantly shifting amalgam of inputs. It is complexly sensitized to a large extent (Dekker & Woods, 1999b; Klein, 1993; Weick, 1993), something that ergonomics has attempted to capture with, for example, intuition and recognition-primed decision models (Klein, 1993, 1998) and a large literature on the nature of expertise (Farrington-Darby & Wilson, 2006). For the work described here, all kinds of patient (i.e., both parturient and fetus) parameters enter into the judgment, ranging from relatively obvious ones such as numbers of previous births, patient weights and conditions, and duration of labor (Drife & Magowan, 2004), to the most subtle physiological signs that only extensive experience can discern. The question of how this constitutes "evidence" remains both contested and deeply problematic in the literature as well as in practice (De Vries & Lemmens, 2006; McDonald, Waring, & Harrison, 2006). Many of the constitutive parameters of X_h remain inexplicit both in clinicians' real-time operational discussions and on reflection—even if any machine output will be assessed in their context. As a result, any approximation of Z_h (after or without a machine output) will remain that at best: a guess by the ergonomist—even before considering any clinician individual differences.

The socially noisy nature of obstetric work accelerates the move from complicated to complex. Like many other parts of medicine, obstetrics is governed both explicitly and implicitly by a relatively rigid medical competence hierarchy: The authority and responsibility for diagnosis and intervention decisions, medication orders, control of medical technology, and continuation of care decisions rests at the top (Ödegård, 2007). This apex is populated by doctors (often male, even in obstetrics), who are often recruited from a limited socioeconomic slice. Underneath lies nursing, which monitors patient condition, carries out medication orders, and offers patient continuity of care (since doctors often only "visit" a patient) (Benner, Malloch, & Sheets,

2010; Ehrenreich & English, 1973). Below that is caring, which handles physiological (if not psychological) needs of feeding, cleaning, and rehabilitation. And below that is the patient, who is generally assumed not to know much of value about his or her disease or condition (Ödegård, 2007).

This strict hierarchy makes each layer subordinate to the one above, which can lead to intriguing divergences between expertise and decision authority. In obstetrics, midwives occupy an important swath of clinical experience and judgment (Sibley, Sipe, & Koblinsky, 2004). Now fully registered nurses with extra training and education, midwives accumulate experience from hundreds or thousands of hours spent by bedsides. Intervention decisions, however, belong formally to those who have not spent such time there. According to both praxis and protocol, doctors only come in when things are no longer normal. But what does that mean, and who gets to say? Signs of "abnormality" are the interpretive and often contested product of the X, the β, and the d' of those present when they occur: the midwives. A physician's intervention decision is thus often preceded by a midwife's intervention decision: to call the doctor.

But midwives typically know the doctors, and doctors often know the midwives. The midwife's setting of β seems to depend on which physician is on duty—estimates of individual physician age, experience, competence, and sometimes even gender affect how much evidence midwives need to call them, rather independent of the number of machine alarms. Doctors, in turn, accumulate their own experience with midwives' βs. Some are known to call some physicians earlier (this may itself depend on a variety of factors, ranging from time of day to bed load and estimates of physician workload, or, more specifically, estimates of that particular physician's ability to handle workload in a socially and professionally acceptable way), which in turn leads to a different physician β for those midwives.

It becomes more intricate still because a doctor on duty can call a doctor on call (who may be elsewhere in the hospital or at home). Midwives, under the rules in the hospitals studied for this chapter, cannot call the backup physician themselves. Instead, midwives make estimates about the likelihood that a duty doctor would call the backup vis-à-vis their desire that he or she do so (which in turn depended on the duty doctor's perceived attributes) and adjusted the construction of their message to the duty doctor on this basis. In other words, the duty doctor's assumed β for calling the backup and his or her assumed d' (sensitivity to evidence) were used to adjust both the timing and the content of the message delivered. Remarks such as "we hope he'll be smart enough to call the backup if we put it this way" seemed to instantiate a narrativized identity of the physician (McDonald et al., 2006), an identity that, with some force, could be converted into that doctor's assumed β and d'.

These narrations created social spaces in the delivery room for the apprehension, identification, and construction of evidence (Iedema, Flabouris, Grant, & Jorm, 2006), including the data generated by ST waveform technology. Further technological developments in one of the hospitals we studied led to summary monitors (of CTG curves from the various delivery rooms) being placed in the obstetric break room. The arrangement became akin to Foucault's (1977) descriptions of the panopticum prison where the very possibility of observation *any* moment changed inmates' behavior *every* moment. In the obstetric ward, the silent knowledge that others could, unbeknownst to oneself, be watching the very evidence trace on which you would be taking action or not was enough to affect clinicians' β again.

With this latest technological intervention, the boundaries of the complex system were made fuzzier again—where exactly did the delivery room end now? Even if physicians might have liked to take action on some CTG traces shown in the break room, they knew better than to enter some midwives' delivery rooms without a call. On the basis of historical experiences, they had their β tuned extremely high. It seemed here that the very medical competence hierarchy could become renegotiated and in some pockets, inverted through processes of social learning and adaptation, fueled largely through narrations and narrativized identities about colleagues' β and d' or even those of colleagues of colleagues. In healthcare, it is into such a world, into such a complex system, that new technology, policy, or managerial interventions are introduced.

COMPLEXITY AND TECHNOLOGICAL OR MANAGERIAL INTERVENTIONS

Rather than a merely complicated problem, for which the solution lies in optimizing the β and d' of the machine and human monitors, the analysis just discussed shows the *complexity* of the problem. Only in a merely complicated system does the insertion of a machine monitor lead to a better β_h (as indeed promised by the vendor). Such a complicated system is a closed system, not open to social, cultural, and professional perfusion at every level; not open to a summary screen in the break room watched by colleagues—or not. Only in a complicated system can X_a and X_h be well correlated.

In contrast, in a complex system, X_h is hugely complexly sensitized and intractable to map. Not only are there interactions between β_h and β_a, but also there are multiple interdependent β_h's in the delivery room alone as nurses of multiple rankings interact in the consideration and construction of evidence from monitored data traces. This system in turn is open. A part of midwives' β_h's is adjusted on the basis of narrativized assumptions about the various β_h's of other clinicians present somewhere in the system. The boundaries of that system are fuzzy (potentially including any other part of the hospital or people's own homes) and can even include the assumed β_h of others about others' β_h (e.g., duty doctor to doctor on call). The kind of message constructed (and what evidence from the ST waveform technology is used to construct it) also hinges on the narrativized d' of the receiver (How clear do we need to be? How thick is this doctor?). Understanding such a complex system is not about understanding the components per se. It is about understanding the intricate web of relationships they weave, their interconnections and interdependencies, and the constantly changing nature of those as people come and go and technologies are adapted in use.

The use of complexity theory can elucidate some of the difficulty of making accurate predictions with the introduction of new technology or policies or other managerial interventions. Complex systems, unlike complicated ones, are not able to be reduced to their components. Complex systems are better modeled after their relationships than on component behavior. Even so, all consequences of technological interventions are not foreseeable, if anything because a definitive description of a complex system remains elusive. Not only are multiple accounts possible and legitimate (depending on who gets to tell those accounts, i.e., from which local position in the complex system), but the system also constantly changes through its own adaptations, additions, and learning—as does its environment, itself also complex.

COMPLEXITY, WORKAROUNDS, AND COMPLIANCE

Can practitioners in a complex system be compliant? Or, is compliance an essentially meaningless term in complexity? Best practice guidelines exist for an increasing number of tasks in healthcare. For example, interventions in obstetrics are indicated by a number of clinical parameters of both parturient and fetus. Proper reading of the evidence by midwives and appropriate execution of the intervention putatively lead to less fetal distress and injury (Benner et al., 2010). "Red rules" (whose transgression is sanctionable) may even exist; a physician needs to be called in case of an abnormal delivery, for instance. If this were the whole system, enforcing compliance makes sense. If intervening were merely a complicated problem, the solution lies in optimizing, through best practice guidelines, the intervention criteria and sensitivity to evidence of those closest to the obstetric process.

The analysis, however, seems to tell another story. There is a serious set of differences between merely complicated systems and complex systems, and assumptions about standardization, norms, and best practices that may work unproblematically in the former are hugely problematic in the latter. Indeed, where does that leave compliance-based quality interventions? Compliance can be defined only in relation to a norm. Some norms (even if vague or negotiable over time) may be readily agreed on in certain pockets of healthcare, and enforcing compliance there doubtlessly has a role to play in creating safety and efficiency. Not using checklists for certain tasks, for instance, may even be construed as not only unnecessarily risky but also unethical (Gawande, 2010; Pronovost & Vohr, 2010).

Max Weber—famous nineteenth-century sociologist—however, warned how bureaucracies, as formal organizations imbued with legal-rational authority, suffer negative consequences when they take their own model of the world too seriously. It can be little more than an administrative palliative to hope that the world is merely complicated, and that it can therefore be controlled or managed. It means believing that existing structures, guidelines, and policies are the instruments of order, and any deviations from them (violations, workarounds) are instances of disorder—the undesirable dark side of human nature that is best contained by more calls for compliance, more guidelines and rules, and more "accountability"—in healthcare often coincident with sanctions (Wachter & Pronovost, 2009).

What does this mean for workarounds? The seemingly neutral definition says that a workaround is a method for overcoming a problem or limitation in a system. "Working around" invokes the existence of a structure, a standard in which something represents a constraint to the achievement of current goals: a constraint that needs surmounting. It is a term that makes eminent sense relative to a complicated system whose "work" is known and fully described. Workarounds constitute the deviant, if inventive, instances when known methods are inadequate and circumvented. In complexity, none of this makes much sense.

But best practice and compliance and their disordered opposites workarounds and violations are the normative rhetorical commitments that belong to a complicated system whose functioning is, in principle, exhaustively knowable and closed to environmental contingency and for which single best methods can be drawn. They are all misleading, or even meaningless, in a complex system that knows no one best method, that is open to contingency and is continually reshaping itself. In complex systems, order emerges from the constantly changing socially and clinically organized circumstances of work and the local rationality of its practitioners who pursue their goals using their knowledge and understanding of the situation. Universal rules and norms that apply to everybody equally all the time amount to a kind of fundamentalist rational Enlightenment ideal whose epistemological and practical reach we should never overestimate. Complex organizations like modern hospitals swiftly and reliably depart from the rationalist expectations of the Weberian model (Vaughan, 1999).

In complex systems, orders of various kinds exist, but they emerge from the multitude of relationships and interconnections and the resulting ways of working. As illustrated, norms for clinical intervention in complex systems are contextual and contingent, varying with time, technology, and social-clinical composition. As people and technologies come and go and learn about each other, relationships change; thus the system constantly reshapes what counts as normative in all kinds of subtle ways. That does not mean that all these ways are desirable or beneficial to the efficiency of care delivery or even patient safety. Efforts, however, to impose a single norm onto complex practice are, not surprisingly, characterized as colonial patronage—as a totalizing, colonizing form of governance that ignores the social and professional richness of clinical work (Holmes, Roy, & Perron, 2008; McDonald et al., 2006).

NEWTON, COMPONENTS, AND COMPLEXITY

The unreflective use of terms such as *complexity* and *compliance* in medicine is not merely a technical problem that easily leads to the misconstruction of the system we all wish to improve. It is ultimately an ethical problem that can carry its implications far beyond the predictions of the clinician or manager who envisioned some policy or technical improvement. It is, apparently, a problem that is easy to tumble into and difficult to avoid. In the West, two hugely influential historical thinkers, Isaac Newton and René Descartes, have pretty much set the agenda for how we think about truth, about cause and effect—and thus about adverse events, about their causes, and what we should do to prevent them.

These effects have become so ingrained, so subtle, so invisible, so transparent, so taken for granted, that we might not even be aware of them. Yet most of the language we speak, and much of the thinking and work we do in patient safety and the prevention of adverse events, is modeled on their ideas. This is not only bad, of course. The teachings of Newton and Descartes have helped us shape and control our world in ways that would have been unfathomable in the time before they were around. But when a patient dies because an analgesic is accidentally pumped into her body via the intravenous line rather than the epidural one, we also call on Newton and Descartes to help us explain why things went wrong. And the consequences of doing that may not always be so helpful for making progress on safety.

Recall the case of the nurse from Wisconsin briefly mentioned in Chapter 7 (Smetzer, Baker, Byrne, & Cohen, 2010). She was charged with criminal "neglect of a patient causing great bodily harm" in the medication error-related death of a 16-year-old girl during labor. The nurse accidentally administered a bag of epidural analgesia by the intravenous route instead of administering the intended penicillin. The criminal complaint (State of Wisconsin, 2006, p. 3) concluded that

- The child's attending physician and the defendant's nurse supervisor reported that the nurse failed to obtain authorization to remove the lethal chemicals that caused the child's death from a locked storage system.
- The nurse disregarded hospital protocol by failing to scan the bar code on the medication, a process in which the nurse had been fully trained and was cognizant. Had the lethal chemicals been scanned, medical professionals would have been forewarned of its lethality, and the death would have been prevented.
- The nurse disregarded a bright, clearly written warning on the bag containing the lethal chemicals prior to injecting them directly into the child's bloodstream.
- The nurse injected the lethal chemicals into the bloodstream in a rapid fashion, failing to follow the approved rate for any medications that may have been prescribed for the child, in an apparent effort to save time. The rapid introduction of these chemicals dramatically hastened the death of the girl, effectively thwarting any ability to save her life.
- The nurse disregarded hospital protocol and failed to follow professional nursing procedures by not considering the five rights of patients prior to the administration of the lethal chemicals (right patient, right route, right dose, right time, and right medication). The practice at her hospital requires the consideration of these five factors at least three times prior to the administration of any medication, the most important procedure established to prevent putting a patient's life in jeopardy through medication errors.

The investigator also reported going to the coroner's office and reviewing the epidural bag (which had contained the analgesic bupivacaine) used in the incident. He found that the bag was labeled with an oversize, hot-pink label on the front side with bold black writing that stated: CAUTION EPIDURAL. The bag had a second label on the backside that was hot pink in color with bold black writing that stated FOR EPIDURAL ADMINISTRATION ONLY. The bag had a white label with a bar code for use with a computerized medication administration system. The penicillin bag, in contrast, did not have a pink cautionary label. It had two small orange labels, indicating the contents of the bag and the patient's name (State of Wisconsin, 2006, p. 6).

The investigator's visit to the coroner's office and the examination of a silent evidence trail was apparently not followed by a visit to the hospital to obtain any idea of the circumstances under which medication is administered. The nurse, for example, had volunteered to take an extra shift because the hospital had difficulty providing nursing staffing and had had only a few hours of sleep during the previous 37 hours.

The nurse, trying to work through the agony of having made a fatal error, faced potential action against her nursing license and lost her job of 15 years.

The criminal charges meant that she also faced the threat of 6 years in jail and a $25,000 fine. As the Institute for Safe Medication Practices (2006) wrote:

> It is important to keep in mind that there is usually much more to a medication error than what is presented in the media or a criminal complaint. For example, while the criminal complaint alleges that the nurse failed to follow the "five rights" and did not use an available bedside bar-coding system, some of the most safety-minded hospitals across the nation with bar-coding systems have yet to achieve a 100% scanning rate for patients and drug containers.

This incident is similar to a 1998 case involving three nurses in Denver who were indicted for criminally negligent homicide and faced a possible 5 year jail term for their role in the death of a newborn who received IV penicillin G benzathine. At first glance, it appeared to many that disciplinary measures might be warranted in that case. But we found more than 50 deficiencies in the medication use system that contributed to the error. Had even one of them been addressed before the incident, the error would not have happened or would not have reached the infant. Fortunately, in the Denver case, the nurse who stood trial was rightfully acquitted of the charges by a jury of laymen that deliberated for less than an hour.

While there is considerable pressure from the public and the legal system to blame and punish individuals who make fatal errors, filing criminal charges against a healthcare provider who is involved in a medication error is unquestionably egregious and may only serve to drive the reporting of errors underground. The belief that a medication error could lead to felony charges, steep fines, and a jail sentence can also have a chilling effect on the recruitment and retention of healthcare providers—particularly nurses, who are already in short supply.

The story that comes to view when the ideas of Newton and Descartes help us explain a failure like this is familiar. Consider the nurse's actions in this case:

- She should have known that such a dose of analgesic coursing into the body intravenously would pretty much lead to death. In fact, the investigator made her admit this on record. Newton also could have told her so. Newton loved the idea that the universe is predictable. As long as we know the starting conditions and the laws that govern it, which each responsible and competent nurse does for his or her little part of that universe, we can predict what the consequences will be.
- The nurse may have lost situation awareness. That is according to Descartes. To Descartes, reality was a binary place. There is the world out there, and then there is the image of that world in our mind. But because we become tired, distracted, or complacent (or, according to the complaint, negligent), we can lose situation awareness, by which the picture in our mind does not entirely line up anymore with reality or is not as complete. The "CAUTION EPIDURAL" label on the analgesic bag was clearly available in the world, but somehow, because of criminal negligence, that caution failed to make it into the mind of the nurse.
- The analgesic, when pumped in intravenously (and certainly when done as quickly as the criminal complaint asserted), represented an overdose. That is according to Newton. Newton was all about energy: the exchange of one form of energy for another (e.g., chemical into physiological, as in this case; more chemical cause, more physiological effect, like cardiac arrest, then death). The Newtonian bastardization has become that if we want to contain danger, we need to contain energy. Or, we have to carefully control its release (like we do in

anything from jet engines to medication). The Swiss cheese model also is about energy and containing it.

- The cause for the patient's death was easy to find. The effect, after all, was the dead patient. And in Newton's world, causes happen before the effect, and ideally they happen close to the effect. The nurse hooked up the wrong bag because of criminal negligence, and now the patient is dead. We found the cause.
- Oh, and what about the nurse? She was fired and charged as a criminal. That is also according to Newton. If there are really bad effects, there must have been really bad causes. A dead patient means a really bad nurse, much worse than if the patient had survived. So much worse, she has got to be a criminal. We cannot escape Newton even in our thinking about one of the most difficult areas of safety: accountability for the consequences of failure.*

Single-factor, judgmental explanations for complex system failures are not unique to healthcare—they are prevalent in fields from military operations (e.g., Snook, 2000), to road traffic (Tingvall & Lie, 2010), to aviation (Holden, 2009). Much discourse about adverse events in complex systems remains tethered to language such as "chain of events" and "human error" and questions such as "What was *the* cause?" and "*Who* was to blame?" The problem of reverting to condensed, single-factor explanations rather than diffuse and system-level ones has of course been a central preoccupation of accident analysis and forms a source of energy behind the human factors approach, which argues against this simplification.

A brief look at the traditional philosophical-historical and ideological bases for sustained linear thinking about failure in healthcare—beyond the issues of identity and competence discussed in the first chapter—is instructive. *Linear thinking* here refers to a process that follows a chain of causal reasoning from a premise to a single outcome. In contrast, complexity and systems thinking regard an outcome as emerging from a complex network of causal interactions and therefore, not necessarily the result of a single factor. A typically linear, Cartesian-Newtonian analysis of failure makes particular assumptions about the relationship between cause and effect, foreseeability of harm, time reversibility, and the ability to come up with the "true story" of an accident. An acknowledgment of the complex, systemic nature of adverse events in healthcare calls for a different approach.

THE CARTESIAN-NEWTONIAN WORLDVIEW AND ADVERSE EVENTS

The logic behind Newtonian science is easy to formulate, although its implications for how we think about adverse events are subtle and pervasive. Classical mechanics, as formulated by Newton and further developed by Laplace and others, encourages a reductionist, mechanistic methodology and worldview. Many still equate "scientific thinking" with "Newtonian thinking." The mechanistic paradigm is compelling in its simplicity, coherence, and apparent completeness and largely consistent with

---

* In a previous book (Dekker, 2007), I suggested ways of thinking about accountability that do not harm safety.

intuition and common sense. The philosophy of Newtonian science is one of simplicity: The complexity of the world is only apparent, and to deal with it we need to analyze phenomena into their basic components (Heylighen, 1999).

The best-known principle of Newtonian science, formulated before Newton by the philosopher-scientist Descartes, is that of analysis or reductionism. The functioning or nonfunctioning of the whole can be explained by the functioning or nonfunctioning of constituent components. Recall from this chapter that attempts to understand the failure of a complex system in terms of failures or breakages of individual components in it—whether those components are human or machine—is common, even if it is inappropriate for complexity. Actually, the inability to find the broken component often equates with a failed investigation. The investigators of the Trans World Airlines 200 crash off New York called it their search for the "eureka part": the part that would have everybody in the investigation declare that the broken component, the trigger, the original culprit, had been located and could carry the explanatory load of the loss of the entire Boeing 747. But for this crash, the so-called eureka part was never found (Langewiesche, 1998).

Newtonian ontology is materialistic: All phenomena, whether physical, psychological, or social, can be reduced to (or understood in terms of) matter, that is, the movement of physical components inside three-dimensional Euclidean space. The only property that distinguishes particles is where they are in that space. Change, evolution, and indeed accidents can be reduced to the geometrical arrangement (or misalignment) of fundamentally equivalent pieces of matter, whose interactive movements are governed exhaustively by linear laws of motion, of cause and effect. The Newtonian model may have become so pervasive and coincident with scientific thinking that, if analytic reduction to determinate cause-effect relationships cannot be achieved, then the adverse event analysis method, department, or person is not seen as entirely worthy.

The determination of the "cause" or "causes" is of course seen as the most important function of an adverse event investigation and assumes that physical effects can be traced back to physical causes (or a chain of causes and effects). In the Newtonian vision of the world, everything that happens has a definitive, identifiable cause and a definitive effect. As alluded to in Chapter 7, there is also a presumed symmetry between cause and effect (they are equal but opposite). Assumptions about cause-effect symmetry can be seen in what is known as the outcome bias (Fischhoff, 1975). The worse the consequences are, the more any preceding acts are seen as blameworthy (Hugh & Dekker, 2009).

According to Newton's image of the universe, the future of any part of it can be predicted with absolute certainty if its state at any time was known in all details. With enough knowledge of the initial conditions of the particles and the laws that govern their motion, all subsequent events can be foreseen. In other words, if somebody can be shown to have known (or should have known) the initial positions and momentum of the components constituting a system, as well as the forces acting on those components (which are not only external forces but also those determined by the positions of these and other particles), then this person could, in principle, have predicted the further evolution of the system with complete certainty and accuracy.

If such knowledge is in principle attainable, then harmful outcomes are also foreseeable. When people have a duty of care to apply such knowledge in the prediction of the effects of their interventions, it is consistent with the Newtonian model to ask how they failed to foresee the effects. Did they not know the laws governing their part of the universe (i.e., were they incompetent, unknowledgeable)? Did they fail to plot out the possible effects of their actions? Indeed, legal rationality in the determination of negligence follows this feature of the Newtonian model:

> Where there is a duty to exercise care, reasonable care must be taken to avoid acts or omissions which can reasonably be foreseen to be likely to cause harm. If, as a result of a failure to act in this reasonably skillful way, harm is caused, the person whose action caused the harm, is negligent. (Global Aviation Information Network, 2004, p. 6)

In other words, practitioners can be construed as negligent if the person did not avoid actions that could be foreseen to lead to effects—effects that would have been predictable and thereby avoidable if the person had sunk more effort into understanding the starting conditions and the laws governing the subsequent motions of the elements in that Newtonian subuniverse.

The trajectory of a Newtonian system is, through its laws of motion, determinable not only toward the future but also toward the past. Given its present state, we can in principle reverse the evolution to reconstruct any earlier state that it has gone through. Such assumptions give adverse event investigators the confidence that an event sequence can be reconstructed by starting with the outcome and then tracing its causal chain back in time. The notion of *re*construction reaffirms and instantiates Newtonian physics: Knowledge about past events is not original, but merely the result of uncovering a preexisting order. The only thing between an investigator and a good reconstruction are the limits on the accuracy of the representation of what happened. It follows that accuracy can be improved by "better" methods of investigation.

Newton argued that the laws of the world are discoverable and ultimately completely knowable. God created the natural order (although God kept the rulebook hidden from humans), and it was the task of the investigator to discover this hidden order underneath the apparent disorder (Feyerabend, 1993). It follows that the more facts an analyst or investigator collects, the more it leads, inevitably, to a better investigation: a better representation of "what happened." In the limit, this can lead to a perfect, objective representation of the world outside (Heylighen, 1999), or one final (true) story, the one in which there is no gap between external events and their internal representation.

This is of course partly the hope or rhetoric behind root cause analysis (RCA). Those equipped with better methods, particularly those who enjoy greater "objectivity" (i.e., those who have no bias that distorts their perception of the world and who will consider *all* the facts), are better positioned to construct such a true story. Formal, adverse event investigations can enjoy this idea of objectivity and truth— if not in the substance of the story they produce, then at least in the institutional arrangements surrounding its production.

NEWTONIAN RESPONSES TO FAILURE IN COMPLEX SYSTEMS

Together, taken-for-granted assumptions about decomposition, cause-effect symmetry, foreseeability of harm, time reversibility, and completeness of knowledge give rise to a Newtonian analysis. It can be summed as follows:

- To understand a failure of a system, investigators need to search for the failure or malfunctioning of one or more of its components. The relationship between component behavior and system behavior is analytically nonproblematic.
- Causes for effects can always be found because there are no effects without causes. In fact, the larger the effect is, the larger (e.g., the more egregious) the cause must have been.
- If they put in more effort, people can more reliably foresee outcomes. After all, they would have a better understanding of the starting conditions, and they are already supposed to know the laws by which the system behaves (otherwise, they would not be allowed to work in it). With those two in hand, all future system states can be predicted and harmful states be foreseen and avoided.
- An event sequence can be reconstructed by starting with the outcome and tracing its causal chain back in time. Knowledge thus produced about past events is the result of uncovering a preexisting order.
- One official account of what happened is possible and desirable, not only because there is just one preexisting order to be discovered but also because knowledge (or the story) is the mental representation or mirror of that order. The *truest* story is the one in which the gap between external events and internal representation is the smallest. The *true* story is the one in which there is no gap.

These assumptions can remain largely transparent and closed to critique in safety work precisely because they are so self-evident and commonsensical. The way they get retained and reproduced is perhaps akin to what Althusser (1984) called "interpellation." People involved in adverse event analysis may be expected to explain themselves in terms of the dominant assumptions; they will make sense of events using those assumptions; they will then reproduce the existing order in their words and actions. Organizational, institutional, and technological arrangements surrounding their work do not leave plausible alternatives (in fact, they implicitly silence them).

Again, RCA investigators are mandated to find the root causes and as a result, frequently turn out enumerations of broken components as their findings. Technological-analytical support (incident databases, error analysis tools) also emphasizes linear reasoning and the identification of malfunctioning components. Newtonian hegemony in adverse event analysis, then, is maintained not so much by imposition as by interpellation, by the confluences of shared relationships, shared discourses, institutions, and knowledge. Foucault called the practices that produce knowledge and keep knowledge in circulation an *epistemé*: a set of rules and conceptual tools for what counts as factual. Such practices are exclusionary. They function in part to establish distinctions between those statements that will be considered true and those that will be considered false (Foucault, 1980). A sociotechnical Newtonian physics is thus read into events that could yield much more complexly patterned interpretations.

COMPLEXITY AND ITS IMPLICATIONS FOR UNDERSTANDING ADVERSE EVENTS

Analytic (or Cartesian-Newtonian) reduction cannot tell how a number of different things and processes act together when exposed to a number of different influences at the same time. This is complexity, a characteristic of a system. Recall from the discussion in this chapter how complex behavior arises because of the interaction between the components of a system. It asks us to focus not on individual components but on their relationships. The properties of the system emerge as a result of these interactions; they are not contained within individual components. Complex systems generate new structures internally; they are not reliant on an external designer. In reaction to changing conditions in the environment, the system has to adjust some of its internal structure. Complexity is a feature of the system, not of components inside it. What would that mean for our understanding of adverse events in healthcare?

Adverse events should be characterized as emergent properties of complex systems (Hollnagel, 2004). They cannot be predicted on the basis of the constituent parts. Rather, they are one emergent feature of constituent components doing their (normal) work. An adverse event is even possible in a complex organization in which people themselves normally suffer no noteworthy incidents, in which everything looks normal, and everybody is abiding by their local rules, common solutions, or habits (Vaughan, 2005). This means that the behavior of the whole cannot be explained by, and is not mirrored in, the behavior of constituent components.

Nonlinearity means that an infinitesimal change in starting conditions can lead to huge differences later. This sensitive dependence on initial conditions removes proportionality from the relationships between system inputs and outputs. Recall from Chapter 5 the evaluation of damage caused by debris falling off the external tank prior to the fatal 2003 space shuttle *Columbia* flight (Columbia Accident Investigation Board, 2003; Starbuck & Farjoun, 2005). Always under pressure to accommodate tight launch schedules and budget cuts (in part because of a diversion of funds to the international space station), certain problems became seen as maintenance issues rather than flight safety risks. Maintenance issues could be cleared through a nominally simpler bureaucratic process, which allowed quicker shuttle vehicle turnarounds. In the mass of assessments to be made between flights, the effect of foam debris strikes was one. Gradually converting this issue from safety to maintenance was not different from a lot of other risk assessments and decisions that the National Aeronautics and Space Administration (NASA) had to do as one shuttle landed and the next was prepared for flight—one more decision, just like tens of thousands of other decisions. While any such managerial, clinical, or engineering decision can be deemed rational given the local circumstances and the goals, knowledge, and attention of the decision makers, interactive complexity of the system can take it onto unpredictable pathways to hard-to-foresee system outcomes.

This complexity has implications for the ethical load distribution in the aftermath of complex system failure. Consequences cannot form the basis for an assessment of the gravity of the cause (or the quality of the decision leading up to it), something that has been argued in the safety and human factors literature (Orasanu & Martin, 1998). It suggests that everyday organizational decisions, embedded in masses of similar decisions and only subject to special consideration with the wisdom of hindsight, cannot be fairly singled out for purposes of exacting accountability (e.g., through criminalization) because their relationship to the eventual outcome is complex and nonlinear and was probably impossible to foresee (Jensen, 1996).

Practitioners in complex systems are capable of assessing the probabilities, but not the certainties, of particular outcomes. With an outcome in hand, its (presumed) foreseeability becomes obvious, and it may appear as if a decision in fact *determined* an outcome, that it inevitably led up to it (Fischhoff & Beyth, 1975). But knowledge of initial conditions and total knowledge of the laws governing a system (the two Newtonian conditions for assessing foreseeability of harm) are unobtainable in complex systems. That does not mean that such decisions are not singled out in retrospective analyses. That they are is but one consequence of Newtonian thinking: Accidents have typically been modeled as a chain of events. While a particular historical decision can be cast as an "event," it becomes difficult to locate the immediately preceding event that was *its* cause. As a result, a human decision (the human error, or violation) is cast as the aboriginal cause, the root cause (Leveson, 2002).

In contrast to the reconstructability of a Newtonian system, the conditions of a complex system are irreversible. The precise set of conditions that gave rise to the emergence of a particular outcome (e.g., an adverse event) is something that can never be exhaustively reconstructed. Complex systems continually experience change as relationships and connections evolve internally and adapt to their changing environment. Given the open, adaptive nature of complex systems, the system after the adverse event is not the same as the system before—many things will have changed, not only as a result of the outcome but also as a result of the passage of time and openness to the environment.

This also means that any predictive power of retrospective analysis of failure is severely limited (Leveson, 2002). Decisions in healthcare organizations, for example, to the extent that they can be excised and described separate from context at all, were not the single beads strung along some linear cause-effect sequence that they may seem afterward. Complexity argues that they are spawned and suspended in the messy interior of organizational life that influences, buffets, and shapes them in a multitude of ways. Many of these ways are hard to trace retrospectively as they do not follow a documented organizational protocol but rather depend on unwritten routines, implicit expectations, professional judgments, and subtle oral influences on what people deem rational or doable in any given situation (Vaughan, 1999).

*Re*constructing events in a complex system, then, is impossible, primarily as a result of the characteristics of complexity. In contrast, the Newtonian belief that is both instantiated and reproduced in official accident investigations is that there is a world that is objectively available and apprehensible. This epistemological stance represents a kind of aperspectival objectivity. It assumes that investigators are able to take a "view from nowhere" (Nagel, 1992), a value-free, background-free, position-free view that is true. This reaffirms the classical or Newtonian view of nature (an independent world exists to which investigators, with proper methods, can have objective access). It rests on the belief that observer and the observed are separable.

Knowledge is nothing more than a mapping from object to subject. Investigation is not a creative process; it is merely an "uncovering" of distinctions that were already there simply waiting to be observed (Heylighen, 1999).

But take again the idea that a sequence of events precedes an adverse event. Who makes the selection of the "events" and on what basis? The very act of separating important or contributory events from unimportant ones is an act of construction, of the creation of a story, not the reconstruction of a story that was already there, ready to be uncovered. Any sequence of events or list of contributory or causal factors already smuggles a whole array of selection mechanisms and criteria into the supposed "re"construction. There is no objective way of doing this—all these choices are affected, more or less explicitly, by the analyst's background, preferences, experiences, biases, beliefs, and purposes. Events are themselves defined and delimited by the stories with which the analyst configures them and are impossible to imagine outside this selective, exclusionary, narrative fore-structure (Cronon, 1992). Whoever gets to pick those has the power to tell the story, to define the "truth" of the adverse event.

Complexity realizes that the observer is not only the contributor to but also in many cases the creator of the observed (Wallerstein, 1996). Cybernetics introduced this idea to complexity and systems thinking: Knowledge is intrinsically subjective, an imperfect tool used by an intelligent agent to help it achieve its personal goals. Not only does the agent not need an objective reflection of reality but also it can actually never achieve one. Indeed, the agent does not have access to any "external reality"; it can merely sense its inputs, note its outputs (actions), and from the correlations between them induce certain rules or regularities that seem to hold within its environment. Different agents, experiencing different inputs and outputs, will in general induce different correlations and therefore develop a different knowledge of the environment in which they live. There is no objective way to determine whose view is right and whose is wrong since the agents effectively live in different environments (Heylighen, 1999).

Different descriptions of a complex system, then (from the point of view of different agents), decompose the system in different ways (Cilliers, 2005). It follows that the knowledge gained by any description is always relative to the perspective from which the description was made. This does not imply that any description is as good as any other. It is merely the result of the fact that only a limited number of characteristics of the system can be taken into account by any specific description. Although there is no a priori procedure for deciding which description is correct, some descriptions will deliver more interesting results than others. It is not that some complex readings are "truer" in the sense of corresponding more closely to some objective state of affairs (as that would be a Newtonian commitment). Rather, the acknowledgment of complexity in adverse event analysis can lead to a richer understanding; thus it holds the potential to improve safety and helps to expand the ethical response in the aftermath of failure.

A Post-Newtonian Analysis of Adverse Events

Complexity and systems thinking denies the existence of one objectively accessible reality that can, as long as we have accurate methods, arbitrate between what is true and what is false. This has implications for what can be considered ethical in the aftermath of failure (Cilliers, 2005):

- An investigation must gather as much information on the event as possible, notwithstanding the fact that it is impossible to gather "all" the information.
- An investigation can never uncover one true story of what happened. That people have different accounts of what happened in the aftermath of failure should not be seen as somebody being right and somebody being wrong. It may be more ethical to aim for diversity and respect otherness and difference in accounts about what happened as a value in itself. Diversity of narrative can be seen as an enormous source of resilience in complex systems, not as a weakness. The more angles there are, the more there can be to learn.
- An investigation must consider as many of the possible consequences of any finding, conclusion, or recommendation in the aftermath of failure, notwithstanding the fact that it is impossible to consider all the consequences.
- An investigation should make sure that it is possible to revise any conclusion in the wake of failure as soon as it becomes clear that it has flaws, notwithstanding the fact that the conditions of a complex system are irreversible. Even when a conclusion is reversed, some of its consequences (psychological, practical) may remain irreversible.

When adverse events in healthcare are seen as complex phenomena, there is no longer an obvious relationship between the behavior of parts in the system (or their dysfunctioning, e.g., human errors) and system-level outcomes. Instead, system-level behaviors emerge from the multitude of relationships and interconnections deeper inside the system and cannot be reduced to those relationships or interconnections. The selection of causes (or events or contributory factors) is always an act of construction by the investigation. There is no objective way of doing this—all analytical choices are affected, more or less explicitly, by the investigation's own position in a complex system, by its background, preferences, language, experiences, biases, beliefs, and purposes. It can never construct one true story of what happened. Truth lies in diversity, not singularity (Cilliers, 2010).

Investigations that embrace complexity, then, might stop looking for the "causes" of failure or success. Instead, they gather multiple narratives from different perspectives inside the complex system, which give partially overlapping and partially contradictory accounts of how emergent outcomes happen. The complexity perspective dispenses with the notion that there are easy answers to a complex systems event—supposedly within reach of the one with the best method or most objective investigative viewpoint. It allows us to invite more voices into the conversation and to celebrate their diversity and contributions.

Let us return to the case of the 16-year-old patient who died after a nurse accidentally administered a bag of epidural analgesia by the intravenous route instead of the intended penicillin administration. The dominant logic of Newton and Descartes helped turn the nurse into the central culprit, leading to her criminal conviction. But the influence of Newton and Descartes did not stop there. An RCA was done. RCAs are designed to track from the proximal events to the distal causes or indeed the root causes. The idea of a root cause, of course, is Newtonian. Effects cannot occur without a cause that we can trace back and nail down somewhere. And that track has a definitive, determinate end: the root cause or causes that triggered all subsequent events.

The RCA identified four proximate causes of the nurse's error. These were the availability of an epidural medication in the patient's room before it was pre- scribed or needed; the selection of the wrong medication from a table; the fail- ure to place an identification band on the patient, which was required to utilize a point-of-care bar-coding system; and a failure to employ available bar-coding technology to verify the drug before administration.

The RCA then set out to explore why each of these proximate causes hap- pened, working its way from the sharp end of the error to the underlying sys- tem problems that contributed to the error. It found that there was no system for communicating the pain management plan of care for the laboring patient to the nurse responsible for getting the patient ready for an epidural. It also found variable expectations from anesthesia staff regarding patient readiness for an epidural and staff scheduling policies that did not guard against exces- sive fatigue. It cited the interchangeability of tubing used for epidural and intra- venous solutions. And it found only a 50% unitwide compliance rate with scanning medications using available bar-coding technology (Smetzer et al., 2010). The dominant logic of an RCA, of course, keeps bringing people back to the fixable properties of their subsystems. The recommendations stemming from the RCA in this case included designing a system to communicate the anesthesia plan of care, defining patient readiness for an epidural, establishing dedicated anesthesia staff for obstetrics, differentiating between epidural and intravenous medications, designing a quiet zone for preparing medications, establishing maximum work-hour policies for staffing schedules, and rem- edying issues with scanning problematic containers (such as the translucent intravenous bags in the case described here) to improve bar-code scanning compliance rates.

Yet as John Stoop, of Delft University of Technology, Delft, the Netherlands, is fond of saying, there is a difference between explanatory variables and change variables. In other words, there is a difference between those things that can explain why a *particular* event happened and those things we should focus our attention on to make sure that *similar* things do not happen in the future. In complex sys- tems, separating these probably makes great sense. The value (see Chapters 6 and 7) of doing retrospective analyses is limited because of the constant dynamics and unforeseeabilities of complexity: that exact failure, in precisely that sequence, will be unlikely to recur.

What is more, fixing properties of a system that brought this failure into being may actually reverberate somewhere else, at some other time. This is a standard fea- ture of complexity. Strengthening something in one corner can lead to vulnerabilities elsewhere. This happens because making changes to some components in a complex system leads to an explosion in relationships and changes in relationships with other components (see Chapter 6). For example, establishing a dedicated anesthesia staff for obstetrics might mean that an anesthetic staff crunch occurs elsewhere at particular times. Or, maximum work-hour policies (part of the same recommendations) get in trouble in another ward because of shortages of those with competence after the creation of a dedicated staff for obstetrics. The interconnectedness of a complex system makes the fixing of broken parts problematic or at least acutely interesting. The change to a part, after all, hardly ever stays with that part. It reverberates through all kinds of other parts in ways that are sometimes foreseeable, but often not.

The very incident that eventually led to the criminalization of this nurse started with such an improvement. A couple of years before, the hospital administration had noticed how anesthetists were unhappy with the working practices in obstetrics. Women in labor could only have their epidural refilled or changed by an anesthetist; midwives and nurses were not allowed to do so. Epidural analgesia used to be given to patients from containers containing 60 milliliters. These came right from the manufacturer and had a dedicated port for connection to the entry point in the patient's back. The problem was that the containers were not very large and used to run out in the middle of the night. That meant that anesthetists had to be called from elsewhere in the hospital just to change or renew an epidural at times when they might just be able to catch some sleep.

After a staff survey, a proposal was developed by those in anesthesia to have the hospital pharmacy itself fill normal 250-milliliter intravenous bags with epidural analgesia fluid. The pharmacy was happy to do so. The time between epidural changes was more than tripled, and anesthetists became much more content. It may have had positive consequences for the continuity, workflow, and even patient safety in other parts of the hospital. In complexity theory, such a small change (or new initial conditions) creates sensitive dependency, however. When new technology was introduced shortly before the death of the 16-year-old, it became obvious that the translucent intravenous bags were virtually impossible to scan with a bar code scanner (nurses had to find and hold a white piece of paper or cloth behind it to bring out the contrast for the scanner to pick up). But a more sinister sensitive dependency was introduced: the interchangeability of the intravenous bag ports. It was possible now to hook up a bag full of analgesic and let it flow straight into the patient through a normal intravenous connection.

If healthcare really is complex, it should start to live with that. It should start to act as if we really understood what that means. Complexity theory, rather than Newtonian reductionism, is where we should look for directions. That is what it should use to consider complex systems that risk drifting into failure. With the introduction of each new part or layer of defense, technology, procedure, or specialization, there is an explosion of new relationships between parts, layers, and components that spreads out through the system. Complexity theory explains how adverse events emerge from these relationships, even from perfectly normal relationships, where nothing (not even a part) is seen as broken. Recall from Chapter 5 that the drive to make systems reliable, then, also makes them complex—which, paradoxically, can in turn make them less safe. Redundancy or putting in extra barriers or fixing them does not provide any protection against a system safety threat. In fact, it can help create, perpetuate, or even heighten the threat. The introduction of a layer of technology (bar code scanning) for double checking a medication order against a patient ID, for example, introduces new forms of work and complexity (the technology does not work as advertised or hoped, it takes time and attention away from primary tasks, and it calls for new forms of creativity and resourcefulness).

There is something seductive about the Newtonian reflex to go down and in to find the broken part and fix it. We can try to tell professionals to be "more professional," for example, or give them more layers of technology to forestall the sorts of component failures we already know about (only to introduce new error opportunities and pathways to failure). Complexity theory says that if we really want to understand failure in complex systems, that we "go up and out" to explore how things are related to each other and how they are connected to, configured in, and constrained by larger systems of pressures, constraints, and expectations.

We would ask why the nurse in question is at work again this day after hardly a break of a few hours (which she spent trying to sleep in an empty hospital bed). We would find that, on a holiday weekend, she was filling an empty slot created by the medical leave of a colleague. Just below, we would find how the subtle but pressing requests to stay for another shift intersect with cultural, personal, and deontological features of those we make into our nurses—of those whom we *want* to be our nurses, those who somehow incarnate commitment and dedication, those who are the embodiment of the "care" in healthcare.

We could trace such a situation to various managerial, administrative, political, and budgetary motivations of a hospital, which we could link to insurance mercantilism, the commercialization of disease, the demand for a commodification of the prices and products of healthcare. We would want to find how, since Florence Nightingale, nursing has steadily lost status, reward, and attraction, with ranks that are hard to fill, its traditional provision of succor eroded under the relentless industrialization of care, and its role as patient voice, as patient advocate now hollow because there is always the next patient and the next. And if we have the societal courage, we might inquire after the conditions and collective norms that make it plausible at the beginning for a 16-year-old girl among us in the community to be pregnant and in need of hospital care.

If we do not dare to go there to undertake this line of inquiry, then it should be no surprise that the cumulative consequences suddenly emerge one day on the work floor of a busy, understaffed ward in a regional community hospital with a patient screaming in acute, severe pain, demanding that something be done now, *now*. If we tinker only gingerly with the final, marginal technical minutiae at various sharp ends, all of those systemic influences will collect repeatedly to shape what any other caregiver will see as the most rational course of action—no matter how large the label on the intravenous bag.

KEY POINTS

- Although complexity is a defining characteristic of healthcare today, many of its managers and practitioners often act as if it were a merely complicated system. The difference matters. Complicated systems are pretty stable, closed to the environment, and ultimately knowable and controllable and should follow one best method. Complex systems are never fully knowable; they are open to the environment and always changing. Order emerges from the multitude of relationships and interactions between component parts.

- The idea that we can understand healthcare as a merely complicated, linear system comes from the dominance of Cartesian-Newtonian thinking in the West. This suggests that we can understand complexity by breaking it down into its component parts, and that the functioning and malfunctioning of those parts will explain the behavior of the whole. But in complexity, that is not the case.

- Rather than thinking about (adverse) events in healthcare as if we can reconstruct the true story by following a good investigative method, complexity suggests that different descriptions of a complex system decompose the system in different ways. The knowledge gained by any description is always relative to the perspective from which the description was made.

REFERENCES

Althusser, L. (1984). *Essays on ideology*. London: Verso.

Amer-Wåhlin, I., Bergström, J., Wahren, E., & Dekker, S. W. A. (2010). *Escalating obstetrical situations: An organizational approach*. Paper presented at the Swedish Obstetrics and Gynaecology Week May 2010, Visby, Sweden.

Amer-Wahlin, I., & Dekker, S. W. A. (2008). Fetal monitoring—A risky business for the unborn and for clinicians. *Journal of Obstetrics and Gynaecology, 115*(8), 935–937; discussion 1061–1062.

Amer-Wahlin, I., Ingemarsson, I., Marsal, K., & Herbst, A. (2005). Fetal heart rate patterns and ECG ST segment changes preceding metabolic acidaemia at birth. *BJOG: An International Journal of Obstetrics and Gynaecology, 112*(2), 160–165.

Amer-Wahlin, I., Yli, B., & Rosen, K. G. (2009). Changes in the ST-interval segment of the fetal electrocardiogram in relation to acid-base status at birth. *BJOG: An International Journal of Obstetrics and Gynaecology, 116*(8), 1138–1139; author reply 1139–1140.

Benner, P. E., Malloch, K., & Sheets, V. (Eds.). (2010). *Nursing pathways for patient safety*. St. Louis, MO: Mosby Elsevier.

Cilliers, P. (1998). *Complexity and postmodernism: Understanding complex systems*. London: Routledge.

Cilliers, P. (2005). Complexity, deconstruction and relativism. *Theory, Culture and Society, 22*(5), 255–267.

Cilliers, P. (2010). Difference, identity and complexity. *Philosophy Today*, 55–65.

Columbia Accident Investigation Board. (2003). *Report Volume 1, August 2003*. Washington, DC: Author.

Cook, R. I., Potter, S. S., Woods, D. D., & McDonald, J. S. (1991). Evaluating the human engineering of microprocessor-controlled operating room devices. *Journal of Clinical Monitoring, 7*(3), 217–226.

Cook, R. I., & Woods, D. D. (1996a). Adapting to new technology in the operating room. *Human Factors, 38*(4), 593–614.

Cook, R. I., & Woods, D. D. (1996b). Implications of automation surprises in aviation for the future of Total Intravenous Anesthesia (TIVA). *Journal of Clinical Anesthesia, 8*(3), 29S–37S.

Cronon, W. (1992). A place for stories: Nature, history, and narrative. *The Journal of American History, 78*(4), 1347–1376.

Dekker, S. W. A. (2007). *Just culture: Balancing accountability and safety*. Aldershot, UK: Ashgate.

Dekker, S. W. A., & Woods, D. D. (1999a). Automation and its impact on human cognition. In S. W. A. Dekker & E. Hollnagel (Eds.), *Coping with computers in the cockpit* (pp. 7–27). Aldershot, UK: Ashgate.

Dekker, S. W. A., & Woods, D. D. (1999b). To intervene or not to intervene: The dilemma of management by exception. *Cognition, Technology and Work, 1*(2), 86–96.

De Vries, R., & Lemmens, T. (2006). The social and cultural shaping of medical evidence: Case studies from pharmaceutical research and obstetrics. *Social Science and Medicine, 62*(11), 2694–2706.

Drife, J. O., & Magowan, B. (2004). *Clinical obstetrics and gynaecology*. Edinburgh: Saunders.

Ehrenreich, B., & English, D. (1973). *Witches, midwives and nurses: A history of women healers*. London: Publishing Cooperative.

Farrington-Darby, T., & Wilson, J. R. (2006). The nature of expertise: A review. *Applied Ergonomics, 37*, 17–32.

Feyerabend, P. (1993). *Against method* (3rd ed.). London: Verso.

Fischhoff, B. (1975). Hindsight ≠ foresight: The effect of outcome knowledge on judgment under uncertainty. *Journal of Experimental Psychology: Human Perception and Performance, 1*(3), 288–299.

Fischhoff, B., & Beyth, R. (1975). "I knew it would happen." Remembered probabilities of once-future things. *Organizational Behavior and Human Performance, 13*(1), 1–16.

Foucault, M. (1977). *Discipline and punish: The birth of the prison*. New York: Pantheon Books.

Foucault, M. (1980). Truth and power. In C. Gordon (Ed.), *Power/knowledge* (pp. 80–105). Brighton, UK: Harvester.

Gawande, A. (2002). *Complications: A surgeon's notes on an imperfect science*. New York: Picador.

Gawande, A. (2010). *The checklist manifesto: How to get things right*. New York: Metropolitan Books.

Global Aviation Information Network. (2004). *Roadmap to a just culture: Enhancing the safety environment*. Global Aviation Information Network (Group E: Flight Ops/ATC Ops Safety Information Sharing Working Group) Washington, DC.

Heylighen, F. (1999). Causality as distinction conservation: A theory of predictability, reversibility and time order. *Cybernetics and Systems, 20*, 361–384.

Holden, R. J. (2009). People or systems: To blame is human. The fix is to engineer. *Professional Safety*.

Hollnagel, E. (2004). *Barriers and accident prevention*. Aldershot, UK: Ashgate.

Holmes, D., Roy, B., & Perron, A. (2008). The use of postcolonialism in the nursing domain: Colonial patronage, conversion and resistance. *Advances in Nursing Science, 31*(1), 42–51.

Hugh, T. B., & Dekker, S. W. A. (2009). Hindsight bias and outcome bias in the social construction of medical negligence: A review. *Journal of Law and Medicine, 16*(5), 846–857.

Iedema, R., Flabouris, A., Grant, S., & Jorm, C. (2006). Narrativizing errors of care: Critical incident reporting in clinical practice. *Social Science and Medicine, 62*, 134–144.

Institute for Safe Medication Practices. (2006, November 7). *ISMP opposes criminal charges for Wisconsin nurse involved in medication error*. Huntingdon Valley, PA: Author.

Jensen, C. (1996). *No downlink: A dramatic narrative about the* Challenger *accident and our time*. New York: Farrar, Straus, Giroux.

Kerstholt, J. H., Passenier, P. O., Houttuin, K., & Schuffel, H. (1996). The effect of a priori probability and complexity on decision making in a supervisory control task. *Human Factors, 38*(1), 65–78.

Klein, G. A. (1993). A recognition-primed decision (RPD) model of rapid decision making. In G. A. Klein, J. Orasanu, R. Calderwood, & C. E. Zsambok (Eds.), *Decision making in action: Models and methods* (pp. 138–147). Norwood, NJ: Ablex.

Klein, G. A. (1998). *Sources of power: How people make decisions.* Cambridge, MA: MIT Press.

Langewiesche, W. (1998). *Inside the sky: A meditation on flight.* New York: Pantheon Books.

Leveson, N. (2002). *System safety engineering: Back to the future.* Boston: MIT Aeronautics and Astronautics.

McDonald, R., Waring, J., & Harrison, S. (2006). Rules, safety and the narrativization of identity: A hospital operating theatre case study. *Sociology of Health and Illness, 28*(2), 178–202.

Moray, N., & Inagaki, T. (2000). Attention and complacency. *Theoretical Issues in Ergonomics Science, 1*(4), 354–365.

Nagel, T. (1992). *The view from nowhere.* Oxford, UK: Oxford University Press.

Norman, D. A. (1990). The problem with automation: Inappropriate feedback and interaction, not over-automation. *Philosophical Transactions of the Royal Society of London Series B–Biological Sciences, 327*(1241), 585–593.

Ödegård, S. (Ed.). (2007). *I rättvisans namn* [In the name of justice]. Stockholm: Liber.

Orasanu, J. M., & Martin, L. (1998). *Errors in aviation decision making: A factor in accidents and incidents.* Human Error, Safety and Systems Development Workshop (HESSD), April 1998, Seattle, WA. Retrieved February 2008, from http://www.dcs.gla. ac.uk/~johnson/papers/seattle_hessd/judithlynnep

Pellegrino, E. D. (2004). Prevention of medical error: Where professional and organizational ethics meet. In V. A. Sharpe (Ed.), *Accountability: Patient safety and policy reform* (pp. 83–98). Washington, DC: Georgetown University Press.

Perry, S. J., Wears, R. L., & Cook, R. I. (2005). The role of automation in complex system failures. *Journal of Patient Safety, 1*(1), 56–61.

Pisek, P. E., & Greenhalgh, T. (2001). Complexity science: The challenge of complexity in health care. *British Medical Journal, 323*, 625–628.

Pronovost, P. J., & Vohr, E. (2010). *Safe patients, smart hospitals.* New York: Hudson Street Press.

Sarter, N. B., Woods, D. D., & Billings, C. (1997). Automation surprises. In G. Salvendy (Ed.), *Handbook of human factors/ergonomics.* New York: Wiley.

Sibley, L., Sipe, T. A., & Koblinsky, M. (2004). Does traditional birth attendant training improve referral of women with obstetric complications: A review of the evidence. *Social Science and Medicine, 59*(8), 1757–1769.

Smetzer, J., Baker, C., Byrne, F., & Cohen, M. R. (2010). Shaping systems for better behavioral choices: Lessons learned from a fatal medication error. *The Joint Commission Journal on Quality and Patient Safety, 36*(4), 152–163.

Snook, S. A. (2000). *Friendly fire: The accidental shootdown of US Black Hawks over Northern Iraq.* Princeton, NJ: Princeton University Press.

Sorkin, R. D., & Woods, D. D. (1985). Systems with human monitors: A signal detection analysis. *Human-Computer Interaction, 1*(1), 49–75.

Starbuck, W. H., & Farjoun, M. (2005). *Organization at the limit: Lessons from the* Columbia *disaster.* Malden, MA: Blackwell.

State of Wisconsin, Circuit Court Dane County (Filed 6 November 2006). *Criminal Complaint, Case number 2006 CF 2512*, p. 3.

Tingvall, C., & Lie, A. (2010). *The concept of responsibility in road traffic* [Ansvarsbegreppet i vägtrafiken]. Paper presented at the Transportforum, January 2010, Linkoeping, Sweden.

Vaughan, D. (1999). The dark side of organizations: Mistake, misconduct, and disaster. *Annual Review of Sociology, 25*, 271–305.

Vaughan, D. (2005). System effects: On slippery slopes, repeating negative patterns, and learning from mistake? In W. H. Starbuck & M. Farjoun (Eds.), *Organization at the limit: Lessons from the* Columbia *disaster* (pp. 41–59). Malden, MA: Blackwell.

Wachter, R. M., & Pronovost, P. J. (2009). Balancing "no blame" with accountability in patient safety. *New England Journal of Medicine, 361*, 1401–1406.

Wallerstein, I. (1996). *Open the social sciences: Report of the Gulbenkian Commission on the Restructuring of the Social Sciences.* Stanford, CA: Stanford University Press.

Weick, K. E. (1993). The collapse of sensemaking in organizations: The Mann Gulch disaster. *Administrative Science Quarterly, 38*(4), 628–652.

Wiener, E. L. (1988). Cockpit automation. In E. L. Wiener & D. C. Nagel (Eds.), *Human factors in aviation* (pp. 433–462). San Diego, CA: Academic Press.

Woods, D. D., & Dekker, S. W. A. (2000). Anticipating the effects of technological change: A new era of dynamics for human factors. *Theoretical Issues in Ergonomics Science, 1*(3), 272–282.

Zaccaria, A. (2002, November 7). Malpractice insurance crisis in New Jersey. *Atlantic Highlands Herald.* http://www.ahherald.com/physicians_forum/2002/pf021107_malpractice.htm

Index

A

Acceptable performance, boundary of, 132
Accountability, 187–212
    accountability, blame and, 196
    at-risk behavior, 188
    confidential information, 195
    criminalization of medical error, 197–205
        crimes as inherently real or as constructed
            phenomena, 197–203
        concern, 203–205
    culpability, 201
    defensive medicine, 204
    disclosure, 194, 195
    employee fear of termination, 208
    Hippocratic principles, 194
    judicial questions, 199
    just culture, 187–196
        accountability free is not blame free, 196
        achieving organizational justice,
            192–194
        assigning behavior (power, production,
            and protection), 189–192
        assigning behavior (right category),
            188–189
        discretionary space for accountability,
            195–196
        reporting versus disclosure, 194–195
    justice, perception of, 193
    key points, 207–208
    learning from failure, 203
    long-shot approaches, 197
    managers, pressures, 190
    matching symptoms to disease, 189
    medication bar code scanners, 190
    motivation, 196
    negligence, 198
    organizational justice, 192
    overcriminalization, 201
    pedagogical questions, 199
    peer and employee assistance, 206–207
    perils of technology, 204
    poststructuralist theory, 201
    primitive belief, 188
    risk conscious society, 202
    room for maneuvering, 195
    safety cultures, 203
    second victim, 205–207
    seriousness of event, 193
    sociological research into deviance, 201
    theological questions, 199
    training deficiencies, 192
    unconditional sympathy, 206
ADU, *see* Automated dispensing unit
Adverse event(s)
    Cartesian-Newtonian worldview and,
        225–235
        complexity and its implications for
            understanding adverse events,
            229–231
        Newtonian responses to failure in
            complex systems, 228–229
        post-Newtonian analysis of adverse
            events, 231–235
    incompetence and, 101
    interactions causing, 129
    investigations, 54, 151–165
        first and second stories, 151–153
        hindsight bias, 153–155
        reconstruction of human contribution to
            adverse event, 155–160
        results, quantification of, 145
At-risk behavior, 188
Automated dispensing unit (ADU), 87, 93
Automation, *see* New technology and automation
Automation surprises, 86, 91

B

Bad luck, 25
Behaviorism, 38, 59
Bias
    hindsight, 46
    outcome, 51, 226
    pervasiveness, 52–53
Boundary markers, 13
Buggy mental models, 72
Building-block learning, 72, 73
Bureaucratic culture, 108
Burnout, practitioner, 101

C

Call-verify, 179
Caring, process of, 13
Challenger accident, 115, 176
Challenge-response, 179
Cognitive factors, 65–81
    ambiguous cues, 75
    attentional dynamics, 66–71
        cognitive fixation and vagabonding,
            66–68